CLEARING THE AIR IN
LOS ANGELES

CLEARING THE AIR IN LOS ANGELES

The Fight Against Smog

CARL R. OLIVER

Published by The History Press
Charleston, SC
www.historypress.com

Front cover: Smog-shrouded Los Angeles. *Courtesy of the* Herald Examiner *Collection/Los Angeles Public Library*.

First published 2024

Manufactured in the United States

ISBN 9781467156745

Library of Congress Control Number: 2024935304

CONTENTS

PREFACE

One day in 1979, I was standing on a hilltop at Universal City, looking down at part of the city of Los Angeles covered by thick smog. The smog loomed as a huge influence on the city and its people. Research over the next forty-plus years and opportunities to explore the topic with students in my college classrooms led this book to emerge.

Beginning in the late 1940s, *Los Angeles Times* publisher Norman Chandler had his newspaper conduct a public service crusade to inform the public about Los Angeles' smog so as to foster wide understanding of the problem and support for the remedies. This book is offered to further the same purpose.

Clearing the Air in Los Angeles: The Fight Against Smog tells of discovering a never-before-recognized kind of smog and the process used to control it. Much of the story is told in words and pictures provided by people who fought the smog for years throughout the Los Angeles air basin.

ACKNOWLEDGEMENTS

A number of people provided encouragement and significant help during research for this book and preparation of the manuscript. In no particular order, I list: Susan Campbell Bartoletti, Simon Elliott, Molly Haigh, Yuriy Shcherbina, Nahal Mogharabi, Elisa Piccio, Terry Garst, Suzanne Noruschat, Patrick Chandler, William R. Watson, Dave Conrad, Jeff Tarmy, William Acker, Bruce Russell, Bill Garnett, R. John Hoover, Daniel O'Brien, Christina Rice, Melissa Montero, Tricia Gesner, Scott Olsen, Kay Conrad, Loma Karklins, Jean Rickard, Elizabeth Nolan, Jim Parker, Bill Kelly, Mike Harriel, Lupe Salazar, Margaret Brunelle, Maria Haagen-Smit, Carole Prietto, Glen Cass, Hiawatha North, Larry Kolczak, Anneke Simmons and Ross Landry.

Images credited to CC by 4.0 are licensed under Creative Commons License Attribution 4.0 International. See creativecommons.org/licenses/by/4.0.

Excerpts and quotations from issues of the *Los Angeles Times* protected by copyright are included by permission; the *Times* has a long-established history of staunch support for antismog education.

More than half a century has gone by since Los Angeles first discovered it was being attacked by smog. Paper records have turned yellow with age and sometimes disappeared. Archivists sometimes are unable to track historic items' origin or provenance. So, despite the author's due diligence to correctly attribute words and pictures to the proper people, the reader may find what looks like mistaken or even entirely missing attributions. The author is eager to hear any corrections or tips leading to discovery of accurate facts so the story of Los Angeles' historic battle to defeat photochemical smog can be improved. Correspondence can be sent to carl@ethicsprocess.com or PO Box 4888, Thousand Oaks, CA 91359-1888.

Chapter 1

AIR YOU CAN *SEE*!

Not long ago, the city of Los Angeles earned an unofficial, undisputed and uncontested title: Smog Capital of the World—an unwanted title it held unhappily for sixty years. Even today, the title lingers although the original conditions—the thick silver-blue smog attacks—no longer occur.

Those original conditions were memorable. Whether people arrived by driving over the surrounding mountains or by flying into Los Angeles International Airport, their first glimpse of the city was a blanket of shimmering silver-blue haze that lay over the city and obscured the ground.

Smog.

At best, people's reaction would be a sarcastic, "At last, air you can *see*!"

What they could not see, because the haze hid streets and playgrounds, was children. "When I was growing up," an Angeleno told me, "many days the smog was so heavy we were not allowed to play outside at school or at home. The smog was so thick that standing at the front door of my home, I could not see the house across the street." In comments reported in the news media and on the internet, other Angelenos often tell the same life experience. Margaret Fox Brunelle told me the smog on the freeway sometimes was so thick that she couldn't see to drive home from work; she had to pull her car off the road and wait for a while.

Los Angeles was not always that way. Smog snuck up on the city. Generally, in people's minds, July 26, 1943, marked the official start of Los Angeles' smog. Before July 26, no smog. Beginning July 26, smog.

Fourth of July 1929 at Los Angeles' Ocean Park Beach. Sparkling waves, sun-splashed sand and clean, fresh air. *UCLA Special Collections CC by 4.0.*

For years, decades, centuries before July 26, 1943, Los Angeles enjoyed a wonderful reputation as an all-year-round warm, sunny paradise. It was blue skies, heaven dressed in waving palm trees and fragrant groves of orange and lemon trees, a land lit by golden sun shining on glistening white Pacific Ocean beaches washed by sparkling blue waves.

Historically, Los Angeles was built on a coastal floodplain of the Los Angeles River, an attractive spot for people to settle and build a new city. The river carried enough water to support what has been called an impenetrable floodplain forest—tangles of marsh willows and blackberry vines thickly dotted with cottonwoods and oaks, with stands of cattails, bulrushes, pickleweed, cordgrass and leadwort. As the city grew, the river's entire surface flow was captured and piped away for people's use. The riverbed became dry most of the year, but when seasonal rainstorms struck, the river suddenly burst into flood stage while it tried to channel torrents of stormwater toward the Pacific Ocean.[1]

In the first half of the twentieth century, Los Angeles became famous worldwide as Hollywood, motion picture capital of the world, a glamorous city populated by movie stars wearing sunglasses and driving fancy cars. Like a vacation resort, it was muscle builders and spirited bathing beauties playing volleyball on the beach and sunning beside backyard swimming pools. It had a business foundation, too: aircraft factories, steel mills and an abundance of companies large and small.

Unexpectedly, something happened on the morning of Monday, July 26, 1943. Between 7:00 and 8:00 a.m., a never-seen-before shimmering silver-blue haze filled the air downtown and grew so thick that no one could see the city's biggest buildings just four blocks away.

People phoned city hall and the newspapers to complain the haze made their nose, throat and eyes hurt. They called it "haze" or "fumes" in 1943. The label "smog" was not used yet. They could plainly see the haze outdoors. Indoors, it was almost invisible, but noses and throats reacted and tears flowed from eyes. Inside one municipal courtroom, the judge promised to adjourn court if the haze persisted.

Historic photo of July 26, 1943—silver-blue haze in downtown Los Angeles, hiding buildings just four blocks away. Los Angeles Times *Photographic Archive, UCLA Library Special Collections.*

Mrs. Walter Maben, with eight-year-old neighbor Barbara Moody, poses for a news photo documenting the Mabens' decision to move away from Los Angeles' smog. *Associated Press photo.*

About noon, the thick haze disappeared. But during the rest of July, August, September and October, it came back again and again. People called the haze almost unbearable. They wanted to get away from the stinky, silvery-blue stuff. The haze was not air you wanted to breathe. As political theater to make that point, some people donned gas masks. Others threatened to move out of town.

Moving elsewhere was no small challenge, what with jobs located in town, household budget limits and the community of family, friends, neighbors, doctors, schools and sports that people are loathe to abandon. But over time, people did it. William Garnett for one. He won notice for museum-quality photographs exhibited in the Museum of Modern Art in both New

York and San Francisco, the Metropolitan Museum of New York, the J. Paul Getty Museum and the Smithsonian Institution. On the back of one photo, an aerial view he captured of Los Angeles blanketed in smog from the beach to the mountains, Garnett wrote, "This is why I moved away from Los Angeles."[2]

City Councilman Carl Rasmussen publicly warned that Los Angeles would become a "Deserted Village" if the thick haze continued. The city could then change its name to "Lost Angeles." The City Council unanimously demanded that the city's Health Commission investigate what could be done to eliminate the thick haze.

In 1943, and for at least six years thereafter, what caused that thick blue haze in Los Angeles was a mystery. With 20/20 hindsight, there were clues in 1943, but only Sherlock Holmes might have spotted them and followed the dots.

CLUE 1: CLIMATE HISTORY

In 1542, European explorer Juan Rodriguez Cabrillo and his crew sailed up the west coast of what we now call Mexico and California. Modern historians cannot locate the ship's log or other records Cabrillo kept during the voyage but have found what appears to be a summary. It is written almost entirely in the third person, indicating the summary is not an original account. Scholars believe the summary was prepared before 1570 and attribute it to "Juan Paez," who may have been a sixteenth-century historian or, perhaps, a friend of Cabrillo.

Historian Henry R. Wagner's translation of Paez's summary of Cabrillo's journal says:

> *The Sunday following, the 8th* [of October 1542], *they came to the mainland in a large bay, which they named "Baia de los Fumos"* [Bay of the Smokes] *on account of the many smokes they saw there.... The bay is in 35 degrees; it is an excellent harbor and the country is good, with many valleys, plains, and groves of trees. On the following Monday, the 9th, they sailed from the Baia de los Fuegos* [Bay of the Fires].[3]

Historians disagree about exactly which bay Cabrillo visited. Hubert Bancroft and Henry Wagner decided it was San Pedro Bay. Herbert Bolton thought it was Santa Monica Bay, which is on the opposite side of a roughly

Historians don't know what Juan Rodriguez Cabrillo looked like, so artist Ren Wicks used his imagination to design this commemorative U.S. postage stamp. *Juan Rodriguez Cabrillo stamp © U.S. Postal Service®, All Rights Reserved. Used with Permission.*

ten-mile-wide peninsula. Whichever bay Cabrillo was in, he was looking at land that would become Los Angeles. Wagner notes that in Cabrillo's time, the words *fuegos* and *fumos* were commonly used interchangeably.

Exactly what date did Cabrillo visit? October 8? Maybe not. Professor Claudio Esteva-Fabregat, a Spanish historian at the University of Barcelona, reported it was not unusual in Cabrillo's day to have little concern for chronology.[4] Moreover, Cabrillo sailed under the Julian calendar. In 1582, Pope Gregory XIII revised the calendar. The Gregorian calendar skipped ten days in October, making October 15 follow October 4. So Cabrillo's October 8 might be the time of year now called October 19...or thereabouts.

Whichever date, it was well within what became known as "smog season," peak months for smog in Los Angeles. And it was well within "fire season." Throughout recorded history, wild brush fires have afflicted the Southern California region, principally from mid-May through October, months when brush and trees are so dry that lightning or any man-made spark can trigger wildfires that fill the sky with smoke. Some people also have speculated that Cabrillo might have seen smoke generated by grassland fires set deliberately by Indigenous people to drive rabbits toward hunters. Rabbits are common in the area.

The truth is, no one today knows what Cabrillo looked like. No paintings or drawings of Cabrillo made during his lifetime have been located. Posthumous drawings apparently were made by artists who had never seen Cabrillo. When U.S. postmaster general Anthony M. Frank wanted to issue a commemorative stamp honoring Cabrillo, Los Angeles artist Ren Wicks undertook to create it.

Because Cabrillo is said to have been Spanish or Portuguese, Wicks chose a Latino model, photographed him in costumes rented from Western Costumers' warehouse and sketched fifteen preliminary designs, and postal officials chose the one they liked best. So the twenty-nine-cent U.S. stamp issued in 1992 shows an artist's conception of what Cabrillo might have looked like, and Wick designed the stamp to make Cabrillo look his best, to glorify Cabrillo, to make him look like a hero.[5]

CLUE 2: THE COLOR BLUE

In 1869, prominent Victorian-era physicist John Tyndall observed a bluish haze in tubes filled with organic vapors when he exposed them to a beam of intense light. Born in Ireland, Tyndall earned a PhD in Germany and served as a professor at the Royal Institution, London. His research was pure science, not related to Los Angeles. Although he was interested in climate, it is unlikely he ever traveled to Los Angeles. His work was known to a scientist at the California Institute of Technology who many years later suggested Tyndall may have been the first person to (unknowingly) create Los Angeles' silver-blue smog in a laboratory.

CLUE 3: CARS AND TRUCKS

On February 17, 1938, the *Los Angeles Times* editorial page looked to the future by quoting a report to the Los Angeles County Regional Planning Commission by William J. Fox, its chief engineer, who said, "The greatest problem our populace is facing in the future is traffic."

Before automobiles were invented, most people in cities walked to where they were going. People tried to avoid lengthy walks, so traditional cities grew around a radius of about two miles and then grew upward rather than outward. Walking people preferred to live close to the city center where they worked. The suburbs at the outer edge of the city were the least desirable places to live or work.

The automobile made it possible for people to commute considerable distances to work or shop in the city center. The oldest U.S. cities, like New York, Boston and Philadelphia, were slow to decentralize, but younger cities like Los Angeles, Houston and Phoenix escaped tradition by growing strong suburbs. Los Angeles' decentralization grew quickly because the region had large tracts of open land that accommodated new factories and new tracts of single-family homes. Shoppers and commuters used automobiles to travel throughout the Los Angeles metropolitan area, with millions of people in millions of cars moving—or trying to move through traffic jams—everywhere in the community every day.[6]

CLUE 4: OZONE

According to James Zeder, a vice president of Chrysler Corporation, by 1925 Detroit automakers were aware of unusual deterioration of rubber parts caused by high ozone occasionally encountered in Los Angeles. Zeder said that in 1925, ozone was considered a natural phenomenon, just one of many blessings of the wonderful Southern California climate.[7]

CLUE 5: CLIMATE CHANGE

In 1930, Jackson A. Graves, president of the Farmers and Merchants National Bank of Los Angeles, spoke directly to climate change. He had lived in town for fifty-four years and felt the climate was better in the 1870s than in 1930. He wrote that as early as 8:00 a.m. each day, the air in 1930 Los Angeles is "extremely disagreeable; it gives me congestion of the head and nasal passages."[8]

Sixty-year resident and Los Angeles bank president Jackson A. Graves felt Los Angeles' air was disagreeable by 1930 and gave him head and nasal congestion. *Security Pacific National Bank Collection/Los Angeles Public Library.*

CLUE 6: PLANTS

In the late 1930s, at the California Institute of Technology in the Los Angeles suburb of Pasadena, an anomaly was observed during research into plant growth. Oat seedlings showed significant differences in response to growth hormone depending on the time of day. During the early morning, plants were up to ten times more sensitive than during afternoons or early evening. Yet time of day—even just daylight or dark of night—was concealed from the plants. They were grown in cellar rooms where temperature, humidity and light were always kept constant. One variable was that the cellar was ventilated with a constant flow of outside air.[9]

CLUE 7: TIME AND PLACE

Charles L. Senn took charge of investigation into the historic July 26, 1943 haze attack. His first written report opened with the observation that this problem was not just on one day. The Los Angeles Health Department received a few complaints about eye irritation before the twenty-sixth—on July 19, 22, 23, 24, 25—many on the especially bad day, July 26, and then many again after that on July 27 and 28.

With hindsight, it is worth noting that with respect to the July 22 event, the *Los Angeles Times* reported the air downtown reeked, people's eyes smarted

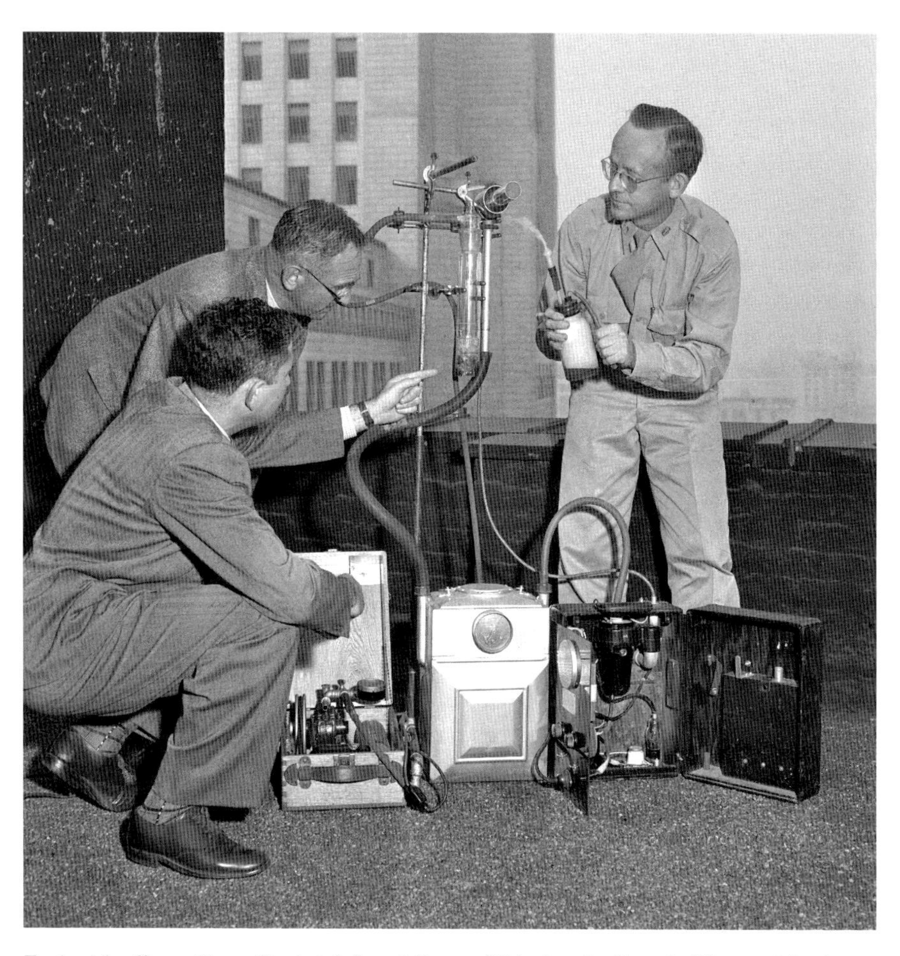

For health officers Harry Kunkel (*left*) and George Uhl, chemist Captain Thomas Marsh demonstrates using an electrostatic precipitator to measure pollutants in Los Angeles' air. *UCLA Special Collections CC by 4.0.*

and their noses were irritated by "acrid ozone." That description may have been overshadowed by an attempt at light humor when the paper also said some people attributed the July 22 attack to a rancorous meeting in the mayor's office about an ongoing streetcar strike.[10]

Senn also reported that this problem was not just downtown. Eye irritation upwind from downtown was equally bad. Most complaints were received from downtown Los Angeles, but some came from the outlying areas of Silverlake, Burbank, Eagle Rock, Pasadena, Alhambra, south to 190th Street, Maywood, Huntington Park and Southgate.

Health inspectors found the problem was not just a 1943 event. Records showed health inspectors had investigated a few eye-burning complaints downtown in previous years: on July 9, 1940, several times during the fall of 1940 and early summer of 1941, on December 10, 1941—a four-hour attack—and on September 21, 1942. They had noted strong effects both downtown and in outlying areas.

On July 27, 1943, Thomas Marsh, a U.S. Public Health Service chemist, used an electrostatic precipitator to collect samples of the hazy air. Analysis showed the eye-irritating air contained aldehydes (strong eye irritants) and aromatic hydrocarbons.

Senn concluded that no single industrial plant could be the source of the eye-irritating fumes for three reasons: The area affected extended at least twenty miles. Fumes and irritation were not especially bad in the vicinity of any one plant. And no employees had reported their company was releasing huge quantities of irritants.[11]

At least seven clues—but no quick solution to the blue haze mystery.

For years, the haze attacked Los Angeles. It got worse and worse and spread into adjacent communities in Orange County, Ventura, Riverside and San Bernardino—the entire Los Angeles air basin.

Scientists encountered many self-proclaimed "experts"—an expert being anyone who had seen, touched, breathed or even heard about the haze—who proposed explanations for its existence and cures that would end it. Political pressure urged scientists to find a solution overnight. Among themselves, scientists found honest disagreement about causes, cures and how to proceed.[12]

The public was impatient, sometimes unreasonable and expected results in a hurry. People joined together in demonstrations urging someone, anyone, to end smog. But the quick cure they dreamed of would not happen. They did not know Los Angeles was about to become the Smog Capital of the World.

Los Angeles Times editorial artist Paul Conrad captured the city's fame for smog in this drawing of an airliner's final approach for landing. *Used with permission of the Conrad Estate.*

Long before the end of 1943, the people of Los Angeles made two things very clear: This thick, silver-blue haze snuck up on them as a complete surprise, and they hated it—this air you can *see*!

Chapter 2

SILVER-BLUE SMOG STRIKES LOS ANGELES

Before the silver-blue haze struck Los Angeles in July 1943, the United States had been actively fighting World War II for more than eighteen months. Every American had been touched by the shock of Japan's attack on Pearl Harbor; the military draft; rationing of everyday necessities like gasoline, tires, sugar, coffee, butter, canned goods and shoes; and bans on using iron and steel in consumer products like batteries, typewriters, thumbtacks and even paperclips. In short, everyone was super-sensitive to the dangers of warfare and the risk of enemy attack.

For Angelenos, that sensitivity had been heightened by local events. Along the California coast, Japanese submarines had skirmished with the U.S. Navy, attacked ships and sunk some merchant ships. In February 1942, a Japanese submarine fired its deck gun at the nearby Ellwood oil fields, only one hundred miles up the beach from Los Angeles, about a two-hour drive.

That submarine later was identified as *I-17*, commanded by Kozo Nishino, who received specific orders to shell any shore target of his own choice about sunset on February 23, 1942, to create panic along the U.S. West Coast. He chose Ellwood, apparently because it had easy access and escape routes for his submarine.

An onshore witness looking through binoculars estimated the submarine was on the surface about one mile offshore at 7:15 p.m. It fired about sixteen shells from its deck gun, an attack lasting about twenty minutes. Reportedly, eleven shells fell short, landing in the ocean. Two landed on ranchland. Three exploded near the Bankline Company oil refinery, which apparently was the intended target, destroying a derrick and a pump house and slightly damaging the Ellwood Pier and a catwalk.

Thereafter, a popular rumor that still circulates to this day said that in 1939, Nishino had been the commander of a Japanese merchant ship, a tanker. His ship stopped at Ellwood to take on a load of oil. Nishino was ashore when he accidentally fell into a prickly-pear cactus and a group of oil workers laughed at him. Nishino felt deeply humiliated. The attack on Ellwood was his revenge.[13]

However, evidence indicates the cactus tale is a myth. Japanese records show Nishino served continuously in the country's navy from 1920, so he never was a merchant seaman. He commanded submarine *I-17* from November 1941 to October 1942 and attacked Ellwood on February 23, 1942. Nevertheless, Japan's psychological goal was at least partially achieved: Southern Californians could feel nervous in February 1942 because Japan had attacked the U.S. mainland for the first time.[14] U.S. Navy intelligence reportedly warned that another Japanese attack could be expected in the next ten hours.

Come evening on February 24, 1942, people in Los Angeles reported seeing flares and blinking lights in the sky. An alert was called at 7:18 p.m. but ended at 10:23 p.m. without incident. After midnight on February 25, radar reportedly picked up an unidentified object 120 miles west of Los Angeles. Antiaircraft batteries went on alert at 2:15 a.m. Radar tracked the object to a few miles from the coast, and the regional controller ordered a blackout at 2:21 a.m.

The object vanished from radar, but reports of enemy planes flowed in. At 2:43 a.m., planes were reported near Long Beach. Minutes later, a coast artillery colonel saw "about 25 planes at 12,000 feet." At 3:06 a.m., four batteries of antiaircraft guns opened fire. Then others began firing. For three hours, swarms of enemy planes were reported.

Later investigation indicated antiaircraft guns' own bursting shells were lit up by searchlights and mistaken for enemy planes. The final tally: no airplanes overhead, no bombs dropped and 1,440 antiaircraft rounds shot down no airplanes. Dawn ended the shooting. The "Battle of Los Angeles" was over. False alarm.[15] Some people reported damage to their homes and cars, but all the damage was attributed to antiaircraft shell debris.

The possibility of enemy attack remained a concern. Los Angeles had grown to be a logical enemy target. Beginning in the last six months of 1940, industry grew fast in Los Angeles to fill orders for the manufacture of war equipment and supplies. When Japan attacked Pearl Harbor on

This *Los Angeles Times* photo documented searchlights hunting for enemy planes suspected to be flying twelve thousand feet over Los Angeles on February 25, 1942. Los Angeles Times *Photographic Archive, UCLA Library Special Collections.*

December 7, 1941, and the United States entered the war as a combatant, Los Angeles already had grown to be "the principal defense manufacturing center of the United States."[16]

So when the unusual thick blue haze struck on July 26, 1943, it was natural for people to wonder if this was a poison gas attack launched by the enemy. People knew poison gas had been used in World War I—horrendous stories were told about it—and in July 1943, officials sought to quickly quell enemy attack rumors.

"Don't Cry! Smarting Gas Not Chemical War" was the headline in the next edition of the *Los Angeles Daily News*, which quoted a city health department spokesperson: "'Tain't tear gas bombs or chemical warfare."[17]

City health officer Dr. George M. Uhl was searching for the true cause. His health inspectors, checking all industrial plants capable of making irritating fumes, already had found a new factory located downtown that was manufacturing a chemical named butadiene and spewing smelly fumes into the air.

Journalists were quick to blame that factory for the haze. It had just opened to produce an ingredient vital to making the synthetic rubber vitally needed to fight the war because Japanese forces had blocked U.S. access to countries exporting natural rubber. Built hurriedly, the factory opened with just the bare-bones equipment needed to make butadiene. Water used in the manufacturing process to wash away impurities flowed into a large open tank, letting smelly fumes escape into the air. Then the water flowed over open-air cooling towers, released additional fumes as it cascaded down and created an airborne spray that people said damaged nearby cars and buildings.

Southern California Gas Company operated the butadiene factory under a contract with the federal government. In a full-page newspaper ad, the company explained its dilemma: shut down the factory to prevent discomfort to the community or keep the factory open to contribute to winning the war.

Newspapers said a factory official wanted to shut down the factory to stop the fumes, but the U.S. Rubber Reserve in Washington, D.C., ordered the factory to continue to make butadiene "regardless of annoyance to the public" because the nation urgently needed synthetic rubber to win World War II.[18]

The U.S. government promised to install a $500,000 "closed system" by December, pipes that would contain the wash water at all times, never expose it to open air and never allow any fumes to escape.[19]

But would that stop the stinky haze? Dr. Uhl was not so sure. Among other things, on July 27, 1943, health inspectors observed that "eye irritation upwind from this plant was as bad as downwind."[20] By marking complaints on a map, inspectors soon saw that eye irritation began simultaneously at points ten miles apart. How could one butadiene factory cause that?[21]

On Monday, October 4, 1943, County General Hospital physicians complained that the haze was harming patients. Factory executive H.L. Masser completely shut down manufacturing operations at the butadiene factory. The next day, even though the factory was still not working, the smelly haze remained so strong that fumes caused great discomfort, if not danger to health, so Superior Court judge Roy V. Rhodes closed his courtroom at city hall.[22]

At the same time, City Councilman John Baumgartner observed about fifty cars stopped on the east side of the butadiene factory because their drivers refused to drive through a visible, thick blanket of fumes lying

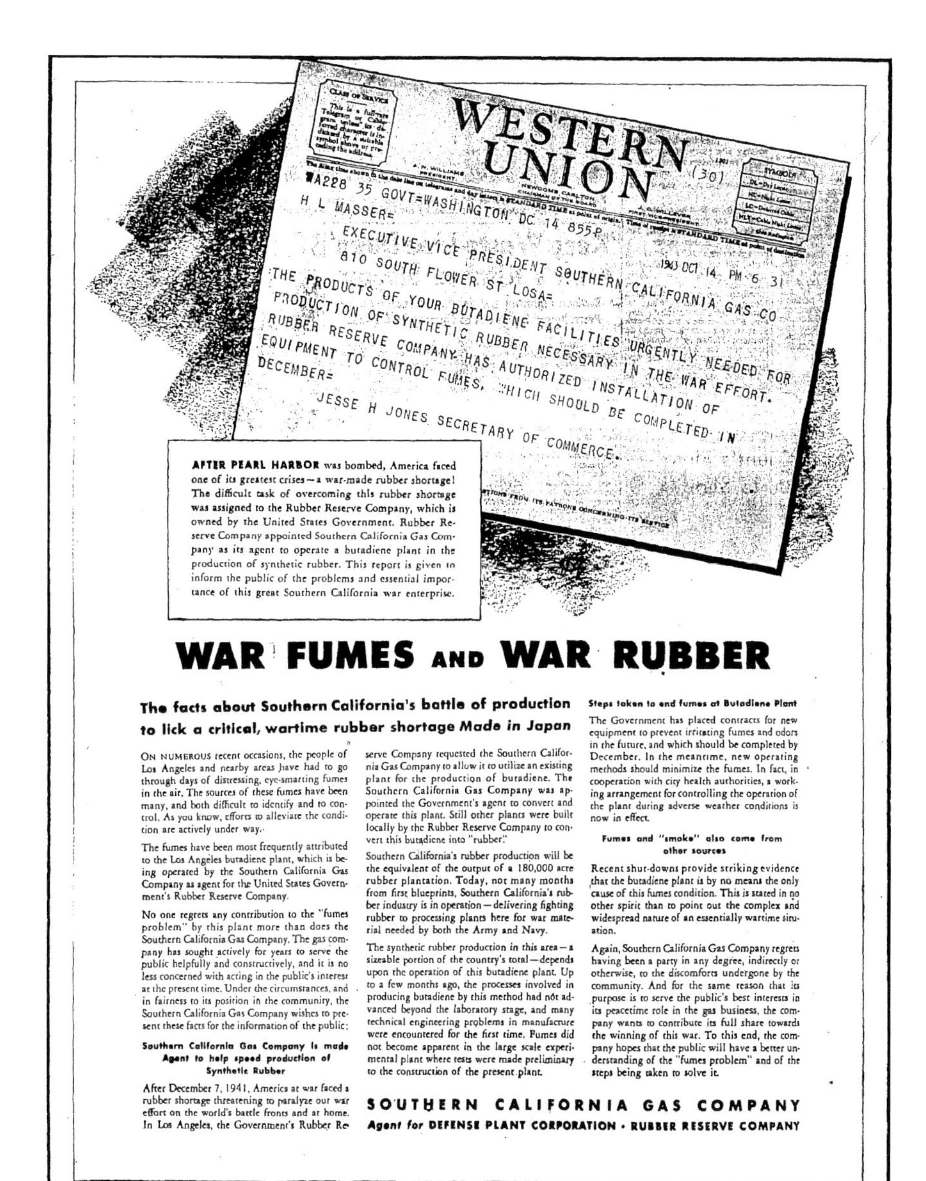

Opposite: Dr. George Uhl wasn't sure closing the butadiene plant would stop the stinky silver-blue haze. *Security Pacific National Bank Collection/Los Angeles Public Library.*

Above: Angelenos wanted the butadiene plant closed, but the U.S. government argued rubber was vital to win the war and promised fumes would end in December. *Southern California Gas Company.*

between the factory and the Civic Center. Baumgartner added that he personally had seen a car usually parked near the factory that was losing its paint due to the fumes. "If the stuff will do that to a car, it can't be very good for the human constitution."[23]

Los Angeles Times reporter Ray Zeman got a bird's-eye view of Los Angeles from the astronomy observatory atop Mount Wilson, home of the then-famous Hooker one-hundred-inch telescope. He reported "huge masses of ugly haze" hid the city's war-booming industrial area below—refineries producing high-octane gasoline, factories producing aircraft, shops creating wood and metal products, or processing food, or building ships, or making rubber.

Zeman interviewed Joseph Hickox, a Mount Wilson astronomer, who said from his bird's-eye perspective atop the mountain and looking down over the entire city of Los Angeles in the coastal floodplain valley below, "It never has been so thick, so muggy, so dirty in all the history of the observatory." Hickox said the haze had increased gradually every summer until 1943, when it intensified noticeably.[24]

Visible air at a busy street corner, silver-blue haze hiding buildings about one block away. Los Angeles Daily News *Collection, UCLA Library Special Collections.*

Torn between his patriotic duty to make butadiene to win the war and his common-sense duty to protect the health of people living in Los Angeles, Masser operated the factory intermittently. The haze was very bad the next Monday, so he shut down the factory for four hours. On Tuesday, he shut it down again at 11:00 a.m. On Wednesday, the city told its attorney to file a lawsuit asking a judge to order the butadiene factory to completely stop operations until fumes-control equipment was installed.

Rubber Reserve official Dr. Henry Cox protested, "Do you want Nazis to take over? The rubber we have made so far is for the war effort."[25]

By October 1943, Los Angeles mayor Fletcher Bowron was heavily involved in the search for a solution to the smog attacks. He notified the city council that "the manager of a large company in the near vicinity of the [butadiene] plant has advised me that employees of various factories and other businesses in the locality are quitting because they object to working under such disagreeable and unhealthful conditions and he expects that it will be very difficult for them to keep sufficient personnel to remain in business."

Moreover, Bowron said he had been told of people hospitalized or suffering due to the fumes and that paint on automobiles standing on the street had been damaged and many cars had to be repainted. "I have done all I can," he said, "and the responsibility now rests with the City Council."[26]

The next day, the city council voted to tell City Attorney Ray Chesebro to take legal steps to immediately close the butadiene plant until devices to prevent release of objectionable fumes were installed. Chesebro filed an injunction lawsuit in federal court.

Washington, D.C., responded quickly. Colonel Bradley Dewey, federal rubber director, flew—a signal of urgency since railroads still dominated long-distance travel—to Los Angeles to speak to city leaders about "the cold facts behind the controversial butadiene plant."

Dewey publicly promised to close the factory every time the haze became offensive. "But," he said, "nobody in Los Angeles has to be told that American civilian economy runs on rubber. We have to stay on wheels. We must get the goods to and from the factories. We need rubber.... The needs of our army, air forces and navy are going up and up beyond the demand that any of us envisioned."[27]

Dewey said, "There is a nuisance here." But the butadiene plant "is the principal source for a group of oil companies in Southern California which are making an appreciable part of the country's synthetic rubber supply." He said when the public knows the facts, when they realize they are helping

to shorten the war and save lives, "I am certain that they will endure these sacrifices until the first of December," the date a mechanical heat exchanger would be completed and installed.

The following day, the city council withdrew its lawsuit. Colonel Dewey thanked them. "You are doing this for the war effort. We, on our part, will see to it that the fume nuisance is abated." Dewey promised the fume nuisance would be eliminated by December 1. "If it isn't, you can call me a bum and I will close down the plant myself."[28]

Opposite: The *Evening Herald and Express* reported Helen Bellinger (*left*) and Carolyn Aberle, with eyes smarting, got ready to battle the city's irritating fumes. Herald-Examiner *Collection/Los Angeles Public Library.*

Right: Colonel Bradley Dewey persuaded Los Angeles to keep the butadiene plant open even if fumes annoyed the public. *Courtesy of University of Southern California, on behalf of the USC Libraries Special Collections.*

Below: The first twenty-one of forty-five closed-pipe water-cooling cylinders installed at the butadiene plant that allowed no fumes to escape into the air. Los Angeles Times *Photographic Archive, UCLA Library Special Collections.*

In November, the butadiene factory closed for six weeks to install closed pipes. It reopened just before Christmas with all its new fumes-control equipment in place. About the same time, the smelly haze seemed to end.

Thus ended 1943, and some people felt victorious.

Trouble was, the smelly haze returned in the summer of 1944. *Los Angeles Times* reporter Ray Zeman reported the city was plagued by more fumes than ever and labeled them "smog (smoke and fog)."[29]

According to the *Oxford English Dictionary*, in London in 1905, Dr. H.A. Des Voeux, treasurer of the Coal Smoke Abatement Society, "did a public service in coining a new word…'smog,' a compound of 'smoke' and 'fog.'" In London, the smoke came from coal commonly burned in the fireplaces, stoves and furnaces of homes and businesses all over the city.

In Los Angeles, coal fires were rare. What's more, Los Angeles did not have the long-lasting, thick fog famous in London. In fact, smog seemed worst on bright, sunny days. At least one scientist studying Los Angeles' smog wanted to change the name. He told me at the time it should be called "smaze—smoke plus haze."

The name smog wrongly guided many people to assume Los Angeles' smog also was smoke plus fog, confusion that led them to be disappointed when reducing or eliminating smoke failed to reduce or eliminate Los Angeles' smog. Research ultimately proved that Los Angeles' smog contains neither smoke nor fog.

Los Angeles Times reporter Ed Ainsworth examined the butadiene factory when its fumes-control equipment was complete and found no smoke or fumes. "Unless this reporter is deaf, dumb and blind and has lost his sense of smell, one of the most-blamed 'sources' of smoke and fumes in Los Angeles is 'not guilty.'"[30]

The big problem remained. Fixing the butadiene factory failed to end Los Angeles' smog.

Chapter 3

A NEWSPAPER LAUNCHES AN ANTISMOG CRUSADE

At the start of 1944, Angeleans knew pollutants were being spewed into the air by 5,500 factories, about 200,000 backyard home incinerators and more than two million people. Dr. Uhl reported successful conferences with business executives at many businesses like railroads, refineries, chemical plants, fish canneries, smelters, electroplaters, fertilizer plants, soap factories, packinghouses and waste disposal plants. He said "voluntary cooperation from plant owners has been good because black smoke indicates inefficiency. It shows good fuel is going up the chimney."[31]

Since the first Stone Age campfire, people likely have complained about smoke polluting the air they breathe. The ancient Romans certainly suffered it. Peter Brimblecombe, lecturer in atmospheric chemistry at the University of East Anglia, Great Britain, found "the poet Horace mentions the blackening of buildings in Rome" and that Seneca, Emperor Nero's tutor, wrote "about the year AD 61 that no sooner did he leave Rome's oppressive fumes and culinary odors than he felt better." In England, the earliest documented air pollution was a 1257 report that Queen Eleanor found the air at Nottingham Castle so polluted with coal smoke that she went to Tutbury Castle to preserve her health.[32]

In *Fumifugium: or The Inconveniencie of the Aer and Smoak of London Dissipated*, John Evelyn praised London as a wonderful city plagued by evil air pollution that

> *renders her less healthy, really offends her, and which darkens and eclipses all her other attributes. And what is all this, but that hellish and dismal*

> *cloud of sea-coal which is not only perpetually imminent over her head… but so universally mixed with the otherwise wholesome and excellent air, that her inhabitants breathe nothing but an impure and thick mist, accompanied with a fuliginous and filthy vapor…corrupting the lungs, and disordering the entire habit of their bodies, so that cathars, phthisics, coughs and consumptions rage more in this one city, than in the whole Earth besides.*

By the 1800s, smoke mixed with fog so frequently covered London that people thought it normal. By 1840, the gray fog of London began to be described as sometimes brown, yellow or orange.

A contemporary description comes from *The Adventure of the Bruce-Partington Plans*, a Sherlock Holmes story written by Sir Arthur Conan Doyle:

> *In the third week of November, in the year 1895, a dense yellow fog settled down upon London. From the Monday to the Thursday I doubt whether it was ever possible from our windows in Baker Street to see the loom of the opposite houses.…For the fourth time, after pushing back our chairs from breakfast we saw the greasy, heavy brown swirl still drifting past us and condensing in oily drops upon the window-panes.*

In the United States, St. Louis, Missouri, won fame not only for its heavy smog but also for its campaign to stop that smog.

The cause of smog in St. Louis was well known. Coal was commonly used by industry and to heat homes. In January 1823, smoke from coal was so dense in parts of St. Louis that people used candles during daylight hours. In 1933, the mayor created a citizen smoke committee and asked Raymond R. Tucker to take charge of improving air quality in St. Louis. Visible smoke emissions from large companies' smokestacks were reduced by two-thirds, but 95 percent of homes still used coal for heat, and therefore, smoke continued to plague the city.

On November 26, 1939, the *Post-Dispatch* newspaper began a campaign to influence people to eliminate visible smoke in St. Louis by switching to fuels like gas, oil, coke or anthracite, which were cleaner but more expensive.

Only two days later, on November 28, 1939, an unprecedented cloud of coal smoke enveloped the city, so black and so thick that the day is remembered as Black Tuesday, what everyone knows as "the day the sun didn't shine." Over the next month, the city experienced nine more days of intense smog.

Left: Raymond R. Tucker, mechanical engineering scholar and politician, built a reputation for eliminating smog in St. Louis, Missouri. *Washington University Archives, St. Louis, Missouri.*

Below: Police train to measure smoke density by comparing it to a Ringelmann chart's circle of dark wedges—20, 40, 60, 80 and 100 percent black. *Courtesy of University of Southern California, on behalf of the USC Libraries Special Collections.*

That experience and the campaign mounted by the *Post-Dispatch* are credited with motivating citizens and the city council to cooperate in abandoning soft coal and switching to cleaner fuels.

By February 1941—the heart of the winter heating season—the *Post-Dispatch* reported smoke and soot so thoroughly eliminated that other cities envied St. Louis. That year, for its antismog campaign, the *Post-Dispatch* won a Pulitzer Prize "for the most disinterested and meritorious public service rendered by any American newspaper during the year," a $500 gold medal.

"St. Louis Did It, Why Can't We?" asked a *Los Angeles Times* headline on September 25, 1944.

Mistakenly, Los Angeles—which had virtually no coal smoke—tried to imitate two cities that fought coal smoke smog, London and St. Louis. Both the city and county of Los Angeles passed laws making it illegal to make any "smoke, dust, soot or fumes" matching or darker than "No. 2 on the Ringelmann Chart" for more than three minutes per hour.[33]

Inspectors were sent into the streets to enforce the law. They checked factory smokestacks and truck exhausts by looking through a hole in the center of their handheld Ringelmann chart and comparing the color of the smoke to five increasingly dark shades printed on wedges around the hole—20, 40, 60, 80 or 100 percent black.

Lack of unity was a complication. Small cities in the Los Angeles air basin ignored the City of Los Angeles and County of Los Angeles laws and continued to allow dense smoke that, of course, flowed across city boundaries without restraint.

"We Can Curb the Smog Nuisance" was the headline of a *Los Angeles Times* editorial on July 20, 1946. It called on the small cities to voluntarily match county anti-air pollution rules; if they refused, then the editorial called on the State of California to impose rules.

People in Los Angeles truly wanted to eliminate the smog, and their efforts successfully reduced visible smoke and fumes. As early as September 1945, inspectors said smoke already was 70 percent less.[34] But the great smog mystery remained unsolved because eye-stinging smog continued as if nothing had happened.

On October 13, 1946, with a feature article written by Ed Ainsworth, the *Los Angeles Times* launched a newspaper crusade to end the smog. The push for that crusade began in Pasadena, a suburb in the heart of the Los Angeles air basin. There, Stephen Royce ran The Huntington, a luxurious resort hotel that attracted extremely wealthy guests year after year.

Left: Stephen W. Royce wanted the *Los Angeles Times* to do something about smog and wanted oil companies to support creating an air pollution control district. *Courtesy of University of Southern California, on behalf of the USC Libraries Special Collections.*

Right: When *Los Angeles Times* publisher Norman Chandler was asked if the newspaper would do anything to end smog, his wife, Dorothy "Buff," answered first: "YES!" *Associated Press photo.*

As Royce told the story, one day a prominent guest said, "Mr. Royce. I am not coming back again next year. I am going to Florida! There is something here in the air which hurts my eyes and burns my throat. I don't know what it is, but I don't like it."

About a week later, another guest said the same thing, and Royce "became really concerned about the future of the hotel."[35]

On Friday, September 13, 1946, smog struck Los Angeles hard, "a dirty gray blanket flung across the city," eye-stinging and sun-dimming.[36] Royce asked *Los Angeles Times* publisher Norman Chandler if the newspaper would do anything to end smog. Chandler's wife, Dorothy Buffum "Buff" Chandler, answered first: a heartfelt "YES!" A month later, the *Times* announced its crusade to end smog. Later, Chandler said his wife deserved credit for goading him into creating the *Times*' antismog campaign.[37] She may have mentioned "something has to be done" more than once; Buff remembered saying it in Norman's office, and he remembered her saying it "sitting next to me in the car."[38]

Chandler tapped forty-four-year-old journalist Edward Maddin Ainsworth to be the face of the campaign. Ainsworth had been a journalist for twenty-eight years, with the *Times* for twenty-two years and in charge of the *Times'* editorial page for the past five years, which was relevant because some readers criticized his antismog campaign articles as sometimes editorial-like biased boosterism, violating the impartiality expected of news reporters. However, the *Times* had a decades-old tradition of editorial boosterism in the spirit of public service, and it pointed out that its campaign sought to focus on facts, not on the sensationalism, exaggerations and scandals yellow journalism used to attract readers and increase circulation.

Reporter Ed Ainsworth led the *Los Angeles Times'* crusade to end the city's silver-blue smog. *UCLA Special Collections CC by 4.0.*

Ainsworth wrote article after article about smog—eighteen in all. The articles blamed smog on smoke and fumes.

> *Then came the war, and with it an overnight industrialization here far beyond the dreams of conventional city planners. New smokestacks belched smoke and fumes. Old smokestacks were overloaded. Great quantities of rubbish were burned. Overtaxed locomotives puffed out their sooty contribution. Diesel engines on trucks, running seven days a week with little time for cleaning and repairs, contributed their viscous black clouds.*[39]

He described Los Angeles air as a cesspool of airborne muck due to natural temperature inversion that exists about two hundred days a year. Normally, air cools as it rises higher. But in Los Angeles, wind blows hot air from the Nevada and Arizona deserts across the top of nearby mountains, and it lies over the city like a warm blanket at an altitude of roughly one thousand feet, sometimes less. Simply, inversion means Los Angeles' air near the ground is colder and therefore heavier than a layer of warm air above it.

A law of physics: cool air cannot rise through warm air. Like a lid, that blanket of warm air traps smoke and fumes in the cold ground-level air, preventing them from rising high into the stratosphere as air circulation normally would carry them.

During an inversion, Los Angeles' air cannot move sideways either. Mountain ranges surround the city like walls on three sides. The fourth side is air over the Pacific Ocean. When wind blows from the ocean toward shore, it creates an invisible fourth wall that pushes air against the mountains. Normally, air pushed against a mountain would slide

In 1945, many rubbish dumps burned trash in Los Angeles County, belching clouds of smoke into the atmosphere. Eventually, all were replaced by cut-and-cover dumps. Herald Examiner *Collection/Los Angeles Public Library.*

up the slope and over the top. But the inversion layer lid blocks upward movement. The air becomes stagnant. None of the pollutants pouring into Los Angeles' air can blow away.

A Los Angeles County Smoke and Fumes Commission report described the everyday muck, saying it is "the smoke and fumes from a thousand [factory smoke] stacks, from tens of thousands of home incinerators, from hundred[s] of thousands of autos and trucks, poisonous vapors from chemical plants, stenches from packing houses [and] sewers and glue factories [and]...refineries and burning rubbish heaps—a hellish cloud that fills the room."[40]

The *Los Angeles Times* printed what people (incorrectly) thought was the recipe for Los Angeles' smog: clouds of smoke meet a low-lying bank of fog, producing smog.[41] That's the recipe for London smog. The stinky haze did wrongly remind some people of the stagnant yellow-brown mixtures of coal smoke and fog fouling London air.

Left: *Los Angeles Times* editorial artist Bruce Russell depicted what people wrongly thought was the recipe for Los Angeles' smog: smoke plus fog. *Used with permission of the Bruce Russell Estate.*

Opposite: By 1957, about one and a half million families in Los Angeles destroyed paper waste daily in smoky trash burner stoves and backyard incinerators. Valley Times *Collection/ Los Angeles Public Library.*

But Los Angeles' smog, at its strongest, was *silver-blue*! What's more, smog struck London hardest on foggy days but struck Los Angeles hardest on sunny days. And Los Angeles' smog caused eyes to burn and tears to flow, something no other city in the world experienced.

When people observed that Los Angeles did not burn much coal, the "answer" was that Los Angeles substituted fumes from thousands of smokestacks, tens of thousands of backyard incinerators, hundreds of thousands of autos and trucks, plus vapors from chemical plants, packinghouses, sewers, glue factories, refineries and burning rubbish heaps.

Raymond Darby, chair of the Los Angeles County Board of Supervisors, described the problem: "We are like the youngster solving the mathematical problem—we know the problem [smog] and can look in the back of the book to see the answer [fresh, clean air]. What lies in between?"[42]

As part of the *Los Angeles Times'* crusade to end smog, Ainsworth pointed to St. Louis as a city that successfully controlled smog. He said that after fifteen years of public education and appeals to cooperate failed, St. Louis defeated smog by making rules and enforcing them—"a vivid object lesson for any city, such as Los Angeles, which also seeks the end of smog."[43]

Chapter 4

SMOG "EXPERT" TUCKER TRIES TO HELP

To climax its antismog crusade and at its own expense, the *Los Angeles Times* hired Professor Raymond R. Tucker to come to Los Angeles to study the city's smog. The *Times* described Tucker as "one of the leading smoke-and-fumes experts of the United States." He served as smoke commissioner of St. Louis from 1937 to 1942, winning recognition as "the man who chiefly was responsible for the tremendous improvement in St. Louis' intolerable smoke situation."[44]

The *Times* reported its initiative received a warm reception from city and county officials. I.A. Deutch, director of the Los Angeles County Office of Air Pollution Control, said he and his staff "are happy that the *Times* has brought an outstanding authority on smoke abatement and combustion engineering to Southern California....Professor Tucker will unquestionably prove of tremendous value to our community by helping to coordinate the various attacks now being made upon the area's No. 1 civic problem."[45]

Building toward Tucker's arrival, during the next week the *Times* printed a series of articles: Municipal Court judge Leo Aggler ruled the city's law prohibiting dense smoke is constitutional; planners recommend closing three public dumps because they use smoky fires to burn trash; the county attorney is organizing support to obtain a new smoke and fumes abatement district; buses, diesel trucks and oil refineries spout fumes; and the temperature inversion layer holds fumes in the air.

On November 8, 1946, Ainsworth's fifth article was headlined "Smug St. Louis Snickers, Says Smog Here Thicker." He reported that the *St. Louis*

Green trimmings from trees and shrubs smolder in piles at the city dump of Glendale, a major suburb in the Los Angeles air basin. Los Angeles Times *Photographic Archive, UCLA Library Special Collections.*

Globe Democrat newspaper printed a long article—with pictures—headlined "Beautiful, Sunny California, Eh? Los Angeles Now No. 1 'Smog Town.'" That article was written by James Murray, who claimed he could hardly see his typewriter keys for the smog. Ainsworth said the article had been "sent in to the *Times* gleefully by many persons who have seen it."

The following day, November 9, Ainsworth's sixth article was headlined "St. Louis Has Key to Smog." He praised the "remarkable success of that city" and the procedure for elimination of the nuisance, which was "the procedure that must be followed by any community."

On December 8, 1946, the day Tucker reached Los Angeles by train, the *Times* printed a page-one editorial that concluded: "Either we take the menace seriously, and fight it seriously, or we can sit and weep in the gloom. Which shall it be?"

The day Tucker arrived, there was not much smog, nor was there the next day. The weather did not cooperate. Truth be told, there was not much smog from the time Tucker arrived until the time he departed two weeks later on December 22.

On the morning of Tucker's first full day in town, December 9, the *Times'* editorial page was devoted entirely to smog articles, principally written by Los Angeles officials:

- I.A. Deutch, county air pollution control director: Questions about smog that remain unanswered
- William E. Simpson, county district attorney: Problems in enforcing existing anti-pollution laws
- Andrew O. Porter, county attorney: Plans for new antismog legislation
- L.E. Timberlake, city public health and welfare committee chair: Trash control problems
- Harry E. Kunkel, city air pollution control director: Request for identification of specific sources of irritants, including the role of automotive exhaust because two million gallons of gasoline are consumed each day in the Los Angeles area

Upon arriving in Los Angeles, Professor Raymond Tucker saw the city's smog from the roof of the Federal Building with Guard Lieutenant R.M. Kermott. Los Angeles Times *Photographic Archive, UCLA Library Special Collections.*

For two weeks, Professor Tucker visited Los Angeles officials and industrial sites. Apparently, he was welcomed everywhere. Everyone wanted Tucker to solve the smog problem.

The *Times* sponsored several events where Professor Tucker spoke. His theme always was that citizens throughout the Los Angeles basin, whatever city, town or community they live in, must join together to battle smog. They must pass new laws to eliminate smog sources and see that officials enforce those laws.

Tucker was part politician (later elected mayor of St. Louis) and part engineering scholar. As a politician, he helped *Times* publisher Chandler create a citizens smog committee. The *Times* invited fourteen prominent citizens to become the *Times* Smog Advisory Committee. One of the fourteen was Stephen Royce, owner of The Huntington. The committee's official charge was to help government officials carry out an antismog campaign based largely on recommendations Professor Tucker would make.

The committee also was expected to keep the public keyed up and ready to accept sacrifices needed to eliminate smog. Tucker said, "An unceasing fight and an aroused citizenry will be necessary to keep your community's air as clean as you want it."[46]

As a scholar, Tucker tried to see every person and thing related to smog. He collected stacks of reports and scientific data to carry back to St. Louis for analysis. Tucker left Los Angeles by train on December 22. "My one regret," he said, "is that you didn't put on a really bad 'smog day' for me. But I'll take your word that you do have them, from the small samples I saw." Ed Ainsworth watched Professor Tucker walk to a window and sniff the air. "Still not much smog. That's the only way in which your hospitality fell down."[47]

Tucker promised to mail his final report to the *Times* in mid-January. In an editorial, the paper asked citizens to be patient because scholarly study takes time.

> *Some persons have seemed a little disappointed that Prof. Raymond R. Tucker, St. Louis smoke and fumes authority brought here by the* Times *to survey the local smog situation, did not function in the spectacular manner of a Main St. magician.*
>
> *Prof. Tucker waved no wands. He rode in no blimps. He visited no movie studios. He kissed no babies.*
>
> *But he did spend from 12 to 14 hours a day on the essential processes of visiting and studying all possible sources of smoke and fumes; amassing scientific data on weather conditions and air currents; interviewing legal*

Professor Tucker watches United States Weather Bureau chief forecaster A.K. Showalter and his colleague Evelyn McConnell launch a weather balloon at Lockheed weather station. Los Angeles Times *Photographic Archive, UCLA Library Special Collections.*

With Professor Tucker, ten of the fourteen citizens on the *Los Angeles Times* Smog Advisory Committee to carry out a public antismog campaign. Los Angeles Times *Photographic Archive, UCLA Library Special Collections.*

> *authorities and digesting engineering material. Now he has returned to St. Louis to put all this mass of material together, and from it to draw his conclusions and make his comprehensive report to Southern California through the medium of* The Times.
>
> *If the* Times *and Prof. Tucker had cared to indulge in a "yellow journalism" stunt and "put the finger" on certain smoke-and-fumes offenders, and scream that if they were suppressed everything would be all right and the sun would shine again forever, tra la, the news stories undoubtedly would have made much better reading. But that was not the purpose. The purpose was to work out quietly and efficiently a line of attack on the problem. The fireworks will come later after Prof. Tucker's report is rendered and the time comes to put the remedies in effect.*[48]

People eagerly anticipated Tucker's solution to smog. The *Times* said it "hoped, like everybody else…that somehow, some way there might be a shortcut to a solution—something like shutting down one big plant or putting muzzles on bus exhausts or making railroad locomotives wear snoods."[49] For many people, "snood" was not an everyday word. For a locomotive, it would be some kind of bag to capture engine smoke—there was a lot of smoke because in those days the steam locomotives burned large amounts of fossil fuels, like coal or oil.

Over the years, people suggested many "solutions":

- Use propellers to make Los Angeles an artificially windy city.
- Gather all available helicopters to hover over the city every inversion day; one thousand should be enough, but two thousand would be better and only a little noisier.
- Install ten giant windmills to blow ground-level air through the warmer air above.
- Place giant fans atop all buildings at least two stories high and on hilltops; aim them to blow fumes away from the city through mountain passes.
- Have huge exhaust fans suck air through giant tunnels bored through the mountains, an air sewer line that would discharge fumes onto the desert or the ocean.
- "Seed" clouds to artificially create rain and wind.
- Have aircraft lay a dense white smokescreen above the city to block out the sunshine.
- Create air circulation by heating ground-level air all over Los Angeles to one hundred degrees Fahrenheit night and day so it will be warmer and lighter than higher air.
- Make artificial rain to clear the air by dropping salt water from airplanes.
- Create air circulation by having 100,000 airplanes drop 7,250,000 tons of water three times each day to make higher air cooler than ground-level air (the suggester admitted he was not sure where to get so much water or how to launch so many airplanes at once).
- Build a $300 million concrete pipeline—pipes twenty feet in diameter, tall enough to hold a two-story house—to move polluted air from the city to the mountains, where smokestacks would release smoke and fumes above the inversion layer.
- Set fire to many gas wells throughout the city so the up-rushing hot air will smite the underside of the inversion layer and push it skyward.

William Larsen drew special attention for proposing an opposite idea: admit smog is here to stay and enjoy its blessings. Among other things, he said, it helps drivers by reducing sun glare, it makes warm days cooler and cool days warmer, it gives people something interesting to talk about and it generates publicity because people elsewhere talk about Los Angeles.[50]

Seriously?

Professor Tucker went back to St. Louis to study the data and write a report. He warned Los Angeles not to expect miracles. Smog control takes time.[51]

Tucker's report arrived at the *Times* in mid-January 1947, just in time for Sunday editions to announce, "*Times* Expert Offers Smog Plan." The full text was published in the newspaper and as a stand-alone flyer distributed widely free of charge.

In a story by Ed Ainsworth published as the lead article on page one of the same issue that printed Tucker's full text, the *Times* highlighted key points of Tucker's report:

- Forbid rubbish burning at private homes.
- End all burning in public or private dumps.
- Stop diesel trucks when they emit dense smoke; issue citations.
- Require chemical analysis of all smoke and fumes from factory stacks.
- Create a unified smog control district, including all cities, towns and communities in the air basin.
- Give "full powers" to smog control officials.

With hindsight, one key point the *Times* failed to highlight came in the fourth paragraph of Tucker's report:

> *Records of the Weather Bureau over the past 10 years show that the total annual mean visibility was a maximum in 1939. From 1939 until 1943 there was a rapid decline, with a minimum being reached in 1942. This same trend was apparent in the semiannual curves with a minimum reached in the same year. It would appear, therefore, that* something had occurred during this period *to cause this reduction in visibility.* [Emphasis supplied.]

People had wondered what happened in July 1943 to kick off the smog attacks. Tucker documented a new time frame—1939 to 1943. But Tucker did not know what had happened.

If Tucker's report held any surprise, it was that he found no explanation for the smog and no cure for it. What about autos and trucks? Ainsworth reported Tucker "absolves automobiles and buses of any major share in causing eye irritation…although pointing out that they, too, must be considered a contributing factor."[52]

I.A. Deutch wears a dome that allows the injection of various gases to discover why Los Angeles' smog causes eyes to burn and tears to flow. Los Angeles Times *Photographic Archive, UCLA Library Special Collections.*

What remained a mystery to Professor Tucker was what ingredient of Los Angeles' smog brought tears to people's eyes. To the best of his knowledge, Los Angeles was the only place where "eye smarting is characteristic of smog." The fumes of the butadiene plant, he said, proved not to be the cause, because tears continued after the plant stopped emitting fumes. Chemical analysis of the air identified irritants, but not enough to cause tears. Tucker encouraged continued research to identify the gases causing tears—the "lachrymating gases."

In the meantime, Tucker said, the sound strategy was to eliminate "all visible discharge into the atmosphere of dust, fumes and gases....In this necessary reduction of contaminants, it is quite possible that the illusive element causing the irritation of the eyes may be reduced below the threshold of human sensitivity."[53]

One week later, the *Times* announced the real test had begun: "It really is up to the people to demonstrate whether they wish a city with clean air." The solution would be neither quick nor easy. "Perhaps some persons expected Prof. Tucker's report to bring forth a startling short cut for the total elimination of smog in a short time. It, of course, did nothing of the kind."[54]

Tucker recommended more of what Los Angeles already was doing: control all sources of visible smoke, fumes and air pollution; get the State of California to create an air pollution control district with power to create and enforce rules throughout the county, a district like he had in St. Louis to force people to obey air pollution regulations; keep doing research on smog.

Chapter 5

LOUIS McCABE HIRED TO END AIR POLLUTION

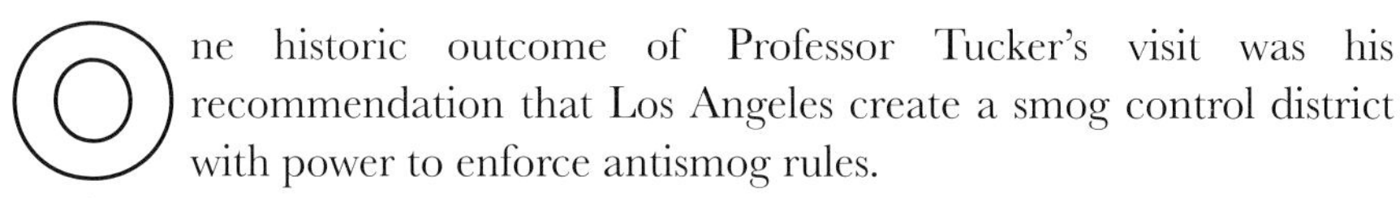

One historic outcome of Professor Tucker's visit was his recommendation that Los Angeles create a smog control district with power to enforce antismog rules.

Tucker proposed it. His personal experience in St. Louis was that such a district could successfully eliminate visible pollutants like coal smoke from the air.

Some people opposed it. How such a district would eliminate Los Angeles' silver-blue smog was unclear. The search for what truly caused the silver-blue smog was unsuccessful so far. Dr. Uhl and his team failed to find it. Professor Tucker failed to find it.

California's legislature almost failed to create it. At the start, lawyers could see that creating such a district for Los Angeles would not be easy because California law made it illegal to give any special district power to *enforce* rules. The solution turned out to be something of a legal shenanigan: instead of a law to create *one* smog control district, the new law created fifty-eight smog control districts—one for each county in California.

Since only one of those counties, Los Angeles, wanted a smog control district, the new law authorized Los Angeles to activate (and pay for) its district while the other fifty-seven counties kept their districts inactive (and cost-free).[55]

Harold Kennedy, an official Los Angeles County attorney, drafted the proposed law. A.I. Stewart, an assemblyman from Pasadena, introduced that proposal as Assembly Bill No. 1. The California Assembly voted for it. The bill would have passed unanimously except for a single "no" cast by Assemblyman

Richard McCollister. His reason? He said the bill declared California air at times is cluttered with dust, smoke, noxious fumes and so on. He did not think legislation would cure it, and why admit it to Florida?[56]

Harold W. Kennedy proposed a law creating fifty-eight smog control districts, one for each county in California, although only Los Angeles County wanted a district. Valley Times *Collection/Los Angeles Public Library.*

Then the bill went to the state senate, where it faced powerful opposition from the oil industry. Stephen Royce, the same man who thought smog drove wealthy guests away from his Pasadena hotel and asked publisher Chandler if the *Los Angeles Times* would do something to end smog, went into action again because he believed the oil companies had enough political clout to block the proposed law.

Royce knew all the top men in all the oil companies very well because they had stayed at San Francisco's famous Fairmont Hotel during the years that he owned it. He had been on hunting trips with them.[57] When he encountered his good friend Charles Jones, president of Richfield Oil Company, Royce said, "Charlie, why are you opposing our smog bill?" Jones' reply was discouraging, to say the least:

> *Steve, I think of you as one of my best friends. I love to hunt and play dominoes with you, but you must realize that I am president of an oil company representing an investment of some $60,000,000. Do you think I am going to allow any smog administrator to tell me whether or not I can run my plant? Not on your life! You wouldn't either if you were in my place.*

Royce asked if Jones would at least try to work out a compromise, "because if you oppose the passing of our law, I know that we can't get it through."[58]

In response, Jones convened a meeting of executives representing all the major oil companies in Southern California, including Standard, Union, Texaco, General Petroleum, Shell and Richfield. He held the meeting in the board of directors conference room at the Richfield Oil Building. The companies were all represented by the president of the company or an executive vice president.

After some "bantering between the oil men as to who had let the cat out of the bag that they were going to oppose Assembly Bill Number 1," the group seriously discussed what they should do about the bill.[59]

The real political break-through came when William Stewart Jr., vice president of Union Oil Company, declared that smog was an important community issue. He said it affected people's comfort and health. It affected the future prosperity of the entire community. The Union Oil Company, he declared, as a responsible citizen of Los Angeles, would not oppose Assembly Bill No. 1.

According to Royce, after Stewart declared support for the new law, no other company's representative "dared to stand up and say that he would oppose."

Royce was ecstatic. "I have always felt that my greatest contribution to Pasadena, outside of the fact that I tried to run a good hotel, was the fight that we made against smog. I am sure that no one realized earlier than I did the harm that it was doing to this area and particularly to the tourist trade."[60]

The major oil companies directed their lobbyist at the state senate, Charlie Stevens, not to oppose the bill. The senate voted unanimously for the new law, and it went to Governor Earl Warren, who signed it on June 10, 1947.

At 10:30 a.m. on the earliest lawful date, October 14, 1947, the Los Angeles County Board of Supervisors activated its new countywide air pollution control district and appointed Dr. Louis Cordell McCabe to be the first director, the first person with legal power to eliminate smoke and fumes everywhere in Los Angeles County.[61] Professor Tucker had recommended creating a countywide air pollution control district. Dr. McCabe was given that district: all of Los Angeles city, all of the unincorporated county area and all of the forty-five other incorporated cities in the county.

Dr. McCabe saw plenty of smog. He arrived in Los Angeles by train on October 1. "As Dr. McCabe stepped from the train," reported Ed Ainsworth in the *Times*, "he was met with a view of the typical gray blanket of fog smoke and fumes which envelops the city so often. 'There is your job!' he was told."[62]

Days later, as he visited smog-plagued neighborhoods, McCabe climbed to the roof of the Green Hotel in Pasadena, at the foot of famous Mount Wilson. Dr. McCabe said, "I've heard rumors that there are mountains around here, but I haven't seen them yet." Smog totally obscured Mount Wilson and all other nearby mountains.[63]

Mayor Bowron closed the city's air pollution control bureau and sent its employees to join McCabe's countywide Air Pollution Control Office.

McCabe's principal strategy: attack *all* sources of visible smoke *all* the time. Inspectors focused on reducing smoke from burning trash dumps, ferrous and nonferrous foundries, steel plants, industrial incinerators, trucks, locomotives and any improper combustion.[64]

Right: "Now I know what my problem is." Dr. Louis McCabe wipes smog tears outside his new office as smog control director, October 3, 1947. Los Angeles Times *Photographic Archive, UCLA Library Special Collections.*

Below: Chairman of the *Times*' Smog Advisory Committee William Jeffers called smoky Kepner Dump a public disgrace and demanded it be closed. *UCLA Special Collections CC by 4.0.*

Although protesters demanded action now, in February 1947, the Los Angeles Board of Supervisors voted 3 to 2 not to close three big burning dumps. *Courtesy of University of Southern California, on behalf of the USC Libraries Special Collections.*

Ed Ainsworth undertook to describe problems of enforcing smog control rules on local companies. Ainsworth identified the worst smoke and fumes violator in Los Angeles as the Los Angeles By-Products Company. "Everybody agrees on that—the owners, the neighbors and the smog control authorities." But the City of Los Angeles had a conflict of interest: the city delivered more than three hundred tons of hard-to-handle trash each day to be burned in the company's furnaces. The company had many air pollution violations, but city officials always had to consider the city's own practical need for this company to continue handling that trash.[65]

The Los Angeles Soap Company's problem was making cheap soap out of recycled household fats. The company installed improvements to control air pollution, but the odor from tanks used to separate glycerin from fats and the dust created in chipping soap into granules both remained problems. Company president F.H. Merrill said the fumes got so bad the company quit making the cheap soap entirely.[66]

At the James Jones Brass Foundry, brass furnaces looked like bottles lying on their sides, revolving slowly with molten metal inside. Gas flames roared through the furnaces, creating a temperature of 2,100 degrees and spewing out smoke and fumes that drifted out openings in the roof.

Owner Wallace Jones Jr. said, "We want to cooperate, but if we put in some kind of blower or washer or something of the kind and then we are told [by smog officials] that it is not satisfactory, we are stuck. It might mean running us out of business."

Jones asked to have the smog engineers review plans in advance and "guarantee that if we install this approved machinery we will not be held responsible after that."[67]

At Apex Steel Corporation, two furnaces heated metal, and the smoke and fumes went up the stack and into the air. After metal was poured, any surplus metal was dumped on the floor, creating both fumes and smoke.[68]

A fifty-two-stack Standard Oil plant spewed "immense quantities" of contaminants into the air—company officials did not deny that. But the plant grew up with Los Angeles and helped contribute to the economic welfare of the community for many years. It helped win World War II and thereafter produced needed fuel and hundreds of other items, including asphalt, cosmetics and candy oils.[69]

The Kennedy Minerals Company was a major offender. But owner John J. Kennedy found a device to eliminate discharge of dust into the air: a series of tall filter cylinders about thirty-five feet high containing filter bags. The dust captured by the filters dropped into bins as fine particles that could be sold for thirty dollars a ton. About one thousand tons were collected each day, so the device was helping pay for itself.[70]

It was difficult to expect wholehearted cooperation from everyone affected by new antismog rules. The public seemed to feel something must be done about smog, but by other people, "not me." Enforcement officers faced pressures to let people follow the procedures they had always used and counter-pressures—official duties—to demand improvements, to stop people from doing what they wanted to do. They faced a public relations challenge, sad stories from individuals and from businesses, nightmares they needed to handle with patience and diplomacy.

And despite their efforts, smog was visibly not going away. On November 22, 1947, a *Los Angeles Times* editorial said, "Some people have begun to ask: 'What's doing about smog? Has the whole thing bogged down?'"

The *Times* defended McCabe, saying it would be fatal to either bog down or rush into ill-advised action. "At this moment Dr. McCabe is working hard on the absolutely essential basic foundations of the antismog program. It is a thankless, unspectacular task." McCabe was setting up standards for smoke and fumes, drafting a permit system, interviewing to hire qualified engineers and technicians—being a good administrator by refusing to act before being ready. After January 1, 1948, said the *Times*, McCabe "will have to begin to show results."

During the next year, McCabe's organized effort visible to the public reduced the number of dumps burning garbage from fifty-seven to fifteen. (Eventually, all were closed and replaced by dig-and-cover dumps.) Two major steel plants were working on emission improvements. Steam locomotives were replaced by diesels. Lumberyards and homes were working to make their incinerators smokeless. (After several years, government outlawed home incinerators and undertook the major task of providing trash pickup for every home in the county.) Refineries were beginning to capture sulfur-laden gases before they escaped into the air. (They processed them into bright-yellow solid sulfur and sold the sulfur for profit.)

However, through the summer and fall of 1948, smog attacks continued and tears still flowed.

Without publicity and unknown to most people, McCabe's team was investing time in another strategy: experiments to identify what pollutants were actually in Los Angeles' smoggy air.

Chapter 6

McCABE'S "BREAKTHROUGH" SCIENCE DISCOVERY

About one year later, after more than five years of searching for the cause of the silver-blue haze now labeled "smog," Los Angeles woke up on Wednesday, September 15, 1948, to news on page one of its morning newspaper, the *Los Angeles Times*, that Dr. Louis C. McCabe, the man it had hired to solve the problem, had successfully found the cause of Los Angeles' smog.

Dr. McCabe had arrived at his new job knowing about London, St. Louis and Pittsburgh smogs. So his strategies undertook to control "all visible and known sources of air pollution" and to scientifically hunt for "invisible smog-forming fumes and gases,"[71] exactly what Professor Tucker recommended.

Many people already suspected the culprit was sulfur dioxide. Like Dr. McCabe, people knew the air pollution villains in London, St. Louis and Pittsburgh smogs were soot from coal smoke and sulfur dioxide. Because soot was virtually nonexistent in Los Angeles, "logic" indicated sulfur dioxide alone must be the problem.[72]

Not so well known, however, was that a visiting science team led by Dr. Edward Weidlein of Mellon Institute, who was widely credited with cleaning up air pollution in Pittsburgh, rather quietly ran tests in Los Angeles that chiefly measured soot and sulfur dioxide. The results indicated Los Angeles' air was cleaner than Pittsburgh's. Baffling![73]

McCabe and his own Los Angeles science staff made what he thought was a breakthrough discovery on August 14, 1948. On that day, McCabe reported, using a cascade impactor they "first collected the droplets (aerosols)

Dr. Louis McCabe (*center*) with Paul Mader (*left*) and Walter Hamming used a cascade impactor to collect droplets McCabe (wrongly) thought were liquid smog. Los Angeles Times *Photographic Archive, UCLA Library Special Collections.*

which constitute the liquid phase of the smog." Chemical tests indicated the liquid was acidic and contained sulfates (sulfur compounds).[74]

McCabe followed up with a confirming test: a flask filled with sulfur dioxide left for twenty minutes in sunlight formed a fog of sulfuric acid. McCabe believed that explained hazy Los Angeles air two or three hours after sunrise when humidity was too low for natural fog to form. That haze likely was sulfuric acid mist.[75]

McCabe said autos were relatively insignificant smog producers. Data showed 822 tons of sulfur dioxide discharged into Los Angeles' air daily, with only 22 tons coming from autos and only 12 tons from diesel engines.

Over at the *Los Angeles Times*, reporter Ed Ainsworth received electron microscope photographs of the "liquid smog droplets" obtained by McCabe's scientists and was pleased, perhaps jubilant, that McCabe's research had at last discovered "the main source of smog in Los Angeles."[76]

The source? Petroleum refineries and chemical plants.

McCabe said refineries, chemical plants and fuel oil released most of the sulfur compounds into Los Angeles' air, about 788 tons each day. A relatively easy solution to the smog problem, McCabe said, would be to open an ammonia manufacturing plant to chemically transform sulfur dioxide into sulfate of ammonia, a useful fertilizer.[77]

But the oil industry protested. Royce's Union Oil Company executive friend W.L. Stewart Jr., speaking for the petroleum industry, was skeptical: "I do not feel that Dr. McCabe's report constitutes scientific demonstration that the oil industry is the sole or even the major contributor to smog in this area."[78]

Two weeks later, after consulting oil company scientists, Stewart said, "We doubt whether the real cause of smog has yet been discovered." He said during the previous two years, refineries already had markedly decreased the amount of sulfur dioxide they spewed into the air by no longer using sulfurous fuel oil and by processing refinery gases to clean hydrogen sulfide from them.[79]

McCabe was harassed. Clean air advocates called him not tough enough, while industry spokespeople called him unreasonable and too tough.[80] Los Angeles' battle to defeat smog always featured such public outcries and political arguing. Now scientific argument joined the fray. The oil companies had hired a nonprofit research firm, Stanford Research Institute (SRI), to find the true cause of smog. SRI cited research findings that smog vapors contain no fewer than eight major components, and still no one knew why smog caused "eye-burn."[81]

McCabe and his team did use their scientific evidence to stop oil refineries from dumping hydrogen sulfide fumes into the air, but smog continued like nothing had happened. Apparently, sulfur compounds were not the primary villains.

Arnold O. Beckman intervened at this point. Beckman had been an assistant professor at the California Institute of Technology. He also was an inventor and entrepreneur. A meter he invented, a science instrument to measure acidity in lemon juice, evolved into an electronic pHmeter that quickly became a vital tool in analytical chemistry. It was the foundation for his creating the National Technical Laboratories, which eventually became the Beckman Instrument Company. His company invented a spectrometer so useful in chemical laboratories that it was called the scientific equivalent of Henry Ford's Model T automobile. The National Inventors Hall of Fame added Beckman to its list of honorees that included Thomas Edison and Alexander Graham Bell.

A behind-the-scenes story that few people know is the important role that Dr. Arnold Beckman played. Beckman did not solve Los Angeles'

Left: Arnold Beckman encouraged his friend Caltech professor Arie Haagen-Smit to analyze Los Angeles' smoggy air using his microchemistry skills and equipment. *Beckman Coulter, Inc.*

Right: Arie and Maria Haagen-Smit. Friends called them by the nicknames Haagy and Zus. *Courtesy of the Archives, California Institute of Technology.*

smog mystery, but he set the force in motion that did solve it: Dr. Arie Haagen-Smit.

Beckman felt uncomfortable with McCabe's conclusion that sulfur caused Los Angeles' smog and wanted to find a better explanation. Sulfur dioxide might be polluting Los Angeles' air, and McCabe had begun a campaign to reduce sulfur dioxide, but Beckman's nose told him McCabe was chasing the wrong smog culprit.

"I knew well the characteristic pungent smell of sulfur dioxide," Beckman said, "and I knew that the human nose can detect sulfur dioxide at extremely small concentrations. I did not smell sulfur dioxide in the air and was therefore reluctant to believe that it was responsible for Los Angeles' smog."[82]

Beckman had been teaching at Caltech in 1938 when he met a new faculty member, Arie Haagen-Smit, known to his friends as Haagy. The university was still struggling financially to survive the Great Depression, and Haagy's paycheck was too small for comfort. About 1940, Beckman helped Haagy find a second job to bring in extra money to support his family. It was a part-time job at the nearby Del Mar horse racing track. Haagy tested the urine of racehorses to detect any drugs given illegally to make them run faster.

Even working two jobs, Haagy was not getting rich. But he saved enough money to buy what his wife, Zus, described as "a nice house" close to Caltech in 1941. So Haagy owed a big favor to his friend Arnold Beckman.

What big return favor could Haagy do for such a man?

It was 1948 before Haagy stumbled onto something. It did not look big at the start. It looked like a small favor. According to Haagy, Beckman simply asked Haagy to attend a meeting. Neither Beckman nor Haagy foresaw how that meeting would change the rest of Haagy's life, how huge that favor would grow.

At the time, Beckman chaired the Scientific Committee of the Los Angeles Chamber of Commerce. The committee members were concerned about polluted air in Los Angeles. What could be causing the pollution? How could the pollution be stopped? Chairman Beckman asked everyone at the meeting to do three things: Sit quietly for a few minutes. Think about the problem. Then write down their ideas and suggestions to solve it.

"I guess I was the only one who wrote," Haagy recalled later. "All I said was, 'Why don't we analyze the air?'"[83]

Chairman Beckman and the committee asked Haagy: Would you please conduct the analysis?

Beckman knew Haagy was the only one at the meeting qualified to analyze Los Angeles' air. He had the education, the experience and the equipment. Air's major components are nitrogen, oxygen and water vapor. Haagy would need to discover what else was in Los Angeles' air.

Actually, Beckman was lucky. Later events proved Haagy was one of the few scientists in the world with enough knowledge and skill to properly do the analysis. Haagy himself downplayed his success, saying if he had not done it successfully, someone else eventually would have.

Haagy's full name was Arie Jan Haagen-Smit, PhD. At the time he died, Haagy had earned the nickname "Father of Smog." (It would be more accurate to call him "Father of Smog Control," and some people did.)

But the truth is that Haagy never wanted to be the world's foremost expert on smog. He did not seek to be a celebrity. His antismog work earned him fame, so much fame that airline flight attendants recognized him and at times he tried to identify himself as simply Mr. Smit, hoping to discourage others' attention.

Haagy was born into a moderately well-to-do family in the Netherlands. His father was the chief chemist for the Netherlands National Mint. Haagy and his two sisters are said to have sometimes played hide-and-seek among the stacks of gold and silver bricks in the mint's storerooms.

In high school, Haagy studied English, French, German, Latin and Greek. He said his only poor grade was in his native language, Dutch. He dreamed of growing up to be a professional mathematician. Mathematics excited Haagy so much that on his own during high school, he taught himself advanced math—calculus.

But Haagy received advice from counselors that led him to change his career plan. The practical problem was that in his hometown, the city of Utrecht, there was only one job for a professional mathematician, a professorship at the University of Utrecht, and it was filled already by a fairly young gentleman who could be expected to keep the job for many years to come. So, when he enrolled as a student at the University of Utrecht, Haagy majored in chemistry because chemists were needed by more employers. Haagy earned a PhD in biochemistry.

In 1930, Haagy married Petronella, a classmate. They had a son, Jan, but Petronella died of lung cancer when Jan was about three months old, leaving Haagy a single father. Haagy arranged to rent a room with board for himself and his son in a private home where the owner would help care for Jan.

In college, both as an undergraduate and graduate student, Haagy had been an athlete, a boxer and a competitive rower. The rowers used a canal that connected the Rhine River with Amsterdam. When he completed his PhD, he was ineligible to be a student rower, so he switched to a civilian rowing group and also became a rowing coach. When he was coaching, by riding his bicycle on the road alongside the rowers' boat in the canal, a young woman named Zus was steering one of the boats. Zus also was one of twelve volunteers helping Haagy with his research, testing the effect of various chemicals on oat seedlings. She began visiting Haagy and Jan at the home they were boarding in.

About eighteen months later, in June 1935, Haagy married Zus, and with Jan, they moved into a rented house on the edge of Utrecht. Haagy became chief assistant to the professor of organic chemistry at the University of Utrecht, the highest position he could hope to earn at that university.

During research in 1934, Haagy identified the plant growth chemical, indole 3 acetic acid. This was a notable discovery. Years later, Haagy's wife, Zus, who assisted that research, described it: "We had little seedlings… maybe an inch tall. We would cut off half of the top and put that chemical on the cut-off half. And if you had the growth hormone, the other half would bend in that direction trying to get some of that growth hormone.... That was the test."[84]

This work, isolating and synthesizing the first recognized plant hormones, aroused the science community's interest in Haagy and led to invitations for him to teach at Harvard and subsequently to join the California Institute of Technology faculty.

In 1935, Haagy received an invitation to teach at Harvard University as a visiting lecturer for the 1936–37 school year. Hitler had started disruptions in Germany in 1933 and gone into Czechoslovakia and Austria, and people saw Europe heading for serious trouble. So moving from the Netherlands to the United States and Harvard looked attractive for more than just career reasons.

Haagy was paid very little at Harvard. A friend rented an apartment for Haagy and his family. They had no car or bicycle. Haagy walked to work. Before Haagy's year at Harvard was up, Dr. Thomas Hunt Morgan invited him to join the faculty at California Institute of Technology as an associate professor of bio-organic chemistry. Frits Went, a former colleague of Haagy's at the University of Utrecht, was already working for Morgan at Caltech and encouraged Haagy to come. Went found an apartment near Caltech for Haagy and his family. Dr. Morgan, the founder of Caltech's biology department, attracted five Dutch professors from Utrecht.

Haagy had never been to Caltech, but he had saved enough money—about $750—to buy a red Ford convertible. Haagy; his wife, Zus; and their two children drove from Harvard to Caltech in three weeks, trying each day to find a motel by 4:00 p.m. so Zus could wash diapers for the next day. "Zus" was a nickname. Her given name was Maria, "which I don't like," she said. *Zus* is Dutch for "Sis," and when she had a choice she always chose "Zus" over "Maria."

At Caltech, Haagy's laboratory became recognized as a U.S. national facility for the specialty of microanalytical chemistry. Haagy collaborated with James Bonner and James English Jr. to identify the plant wound hormone, which they named traumatic acid. This was a major discovery. The hormone helps plants heal when they are cut or otherwise damaged. It stimulates plant cells to form a callus—soft tissue over a cut.

After Haagy's death, his colleague James Bonner spoke of Haagy's early and noteworthy science contribution: isolation of auxin and indole 3 acetic acid as plant-growth material with auxin activity. That "laid the cornerstone of our knowledge of plant growth regulation," Bonner said. And Bonner was

Left: Haagy (*left*) with Thomas Hunt Morgan, founder of Caltech's biology department, who invited Haagy to join the Caltech faculty as associate professor of bio-organic chemistry. *Courtesy of the Archives, California Institute of Technology.*

Right: Frits W. Went encouraged Haagy to join the Caltech faculty. They were two of five Dutch professors from Utrecht who joined the Caltech faculty. *Security Pacific National Bank Collection/Los Angeles Public Library*.

impressed that Haagy claimed no special credit for "such a great discovery." Nor did Haagy claim credit for "an even larger discovery" in 1935, when Haagy and Frits Went found substances never found in nature that mimic indole 3 acetic acid and that created the whole field of selective herbicides.

Haagy investigated the chemistry of many plants, such as pine trees and desert plants, and the chemicals that give flavor and aroma to plants like pineapple and wine grapes. This was scientifically important work that his colleagues believed could eventually be rewarded with a Nobel Prize.

Haagy was thriving and successful at Caltech. What could go wrong?

Chapter 7

HAAGY'S "JACKPOT" EXPERIMENT

As a specialist in microchemistry, Haagy was one of the few people in the world with the experience and equipment needed to properly analyze Los Angeles' air. For other chemists, "small samples" meant 200 to 300 milligrams of substances. For microchemistry, Haagy was used to testing *tiny* samples, 1 milligram or less.[85] A colleague commented that Haagy was excited about seeking exceedingly small quantities of constituents such as those that cause growth, flavor or odors—he was always searching for the grain of salt in the sugar bowl.[86]

At Harvard, Haagy studied a bacterium that grew in hay, so he had large amounts of hay in the basement of one of the botany buildings. At Caltech, Haagy studied the aromas and flavors of radishes, beans, cashews, onions, garlic, marijuana, pine trees and wine grapes.

When Haagy was asked to study Los Angeles' smog, he had been studying pineapple for the Pineapple Research Institute of Hawaii, which had flown three tons of pineapple to Caltech from Hawaii. Haagy's laboratory equipment evaporated the pineapple juice and successfully distilled a few grams of solution containing the chemicals that give pineapple its aroma and flavor.

In coordination with Arnold Beckman, Dr. McCabe and Dr. Robert Vivian, who was head of the chemistry department at the University of Southern California, Haagy undertook to perform a proper microchemistry analysis of Los Angeles' polluted air.[87] He used the same equipment to analyze smoggy air that he had been using to analyze the aroma and flavor

Frank Hirosawa and Arie Haagen-Smit with equipment Haagy said he used to study pineapple and smog. Haagy called smog "the bouquet of Los Angeles." *Courtesy of the Archives, California Institute of Technology.*

of pineapple. "I just opened the window, stuck a pipe out, and drew the air through my equipment."[88]

Haagy pumped that smoggy air through a series of glass pipes and tubes. Ice and salt chilled one tube to minus 20 degrees Celsius; dry ice and acetone chilled a second tube to minus 80 degrees Celsius; and liquid nitrogen chilled

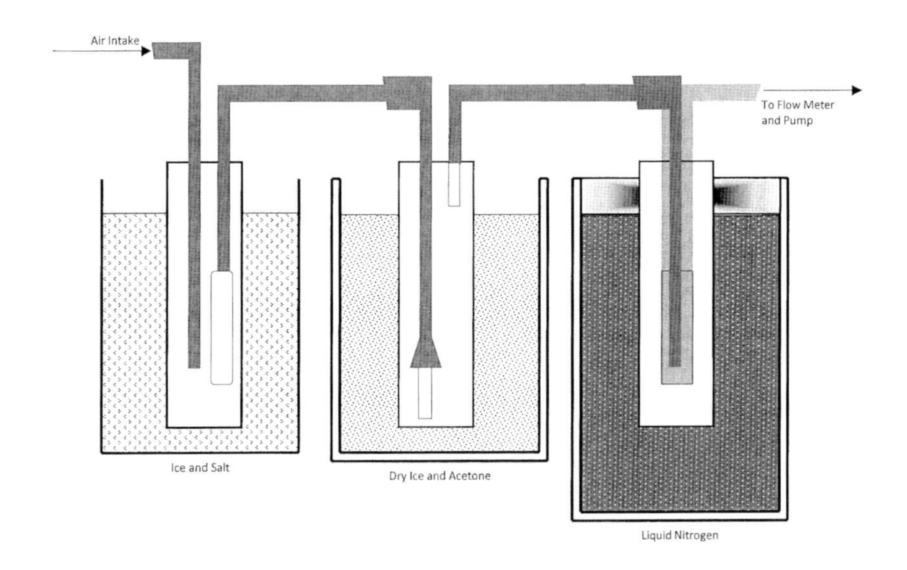

Diagram of the freeze-out procedure Haagy used. Beaker 1 cooled air to minus 20 degrees Celsius, Beaker 2 to minus 80 degrees Celsius and Beaker 3 to minus 181 degrees Celsius. *Author's collection.*

a third tube to minus 181 degrees Celsius. This squeezed from the smoggy air just "a couple drops of dark brown, vile-smelling liquid."[89] It was not a sulfur compound. It was volatile organic peroxides, substances that, among other things, could irritate eyes.

Haagy realized that Los Angeles had a new kind of smog, something the world had never before discovered. Chemically, it was the opposite of coal smoke smog. The coal smoke smog in London, St. Louis and Pittsburgh was *reducing*—it burned oxygen. The chemistry of the never-seen-before smog in Los Angeles was *oxidizing*—the chemistry of volatile organic peroxides *created* oxygen, the unstable and very active form of oxygen called ozone.[90]

That fit! Haagy's nose could smell ozone in Los Angeles' air.[91] Haagy trusted his nose. He considered the nose "a most astonishing chemical laboratory" that in a fraction of a second could identify chemicals it would take a chemist days to analyze in the laboratory.[92] Haagy observed that the nose can easily detect one thousandth of a milligram of ethyl mercaptan (a relative of the skunk smell) released into a small auditorium. For skatol (found in feces), you could chop that amount into ninety-two pieces and smell just one piece.

Haagy knew how to make ozone in the laboratory: shine strong light on air containing traces of hydrocarbons and nitrogen oxides. He surmised nature might use that recipe to make Los Angeles' smog. Los Angeles' famous sunshine would provide plenty of strong light, gasoline fumes would provide traces of hydrocarbons and any kind of burning would create traces of nitrogen oxides. That could create ozone, which might react with additional gasoline fumes to create smog.

In his laboratory, Haagy tested his theory by pouring gasoline into a beaker and blowing ozone over it. The room filled with smog—"the identical eye-irritating, throat-rasping, acrid smog that Angelenos had been breathing for years."[93]

It was a lucky experiment.

Later, Haagy found out just how lucky. He discovered smog formed only when hydrocarbons and nitrogen oxides were present in exactly the right proportions. Too much of something, no smog; too little of something, no smog. *Really* small amounts were involved—three to ten parts per million. Fewer than three, no smog; more than ten, no smog.[94]

Haagy demonstrates making Los Angeles' smog in a Caltech classroom: in strong light, pour gasoline into a beaker and blow ozone over it. *Courtesy of the Archives, California Institute of Technology.*

With some awe, he said, "I hit the jackpot with the first experiment."[95]

After completing his research, extracting the few vile drops and analyzing them, Haagy thought he was finished with smog. "I was working on other things, you know?"[96] He had more than once said he was not interested in studying air pollution and hunting for things people's activities put into the air. He wanted to study curiosities created by nature—the flavors and aromas of food plants as an obvious example.

Louis McCabe faced a serious problem. He had announced that Haagy found the true cause of Los Angeles' smog, but soon notable scientists called that untrue, a false victory.

Haagy's discovery turned into a low point in the search for the cause of and cure for Los Angeles' smog. In science, duplicating experiments is called "replication." Other scientists try to verify discoveries by doing exactly the same experiment. They should get exactly the same results. But other scientists attempting to replicate Haagy's experiments got different results. No respected scientist could make Los Angeles' smog in other laboratories.[97]

Haagy looked like a liar.

Arnold Beckman interceded again. Beckman persuaded Haagy to go with him to hear a talk on air pollution being given at Caltech by a scientist from the Stanford Research Institute (SRI). Actually, Beckman admitted years later that he did more than that. He knew the scientist had criticized Haagy, said Haagy was "all wet" because from the scientist's personal experience the peroxy compounds did not smell, did not irritate the eyes and so forth. So Beckman "got [the scientist] to give a talk over here [at Caltech] in the chemistry lecture room, and I got Haagy to come and sit with me."

The scientist described his own work and criticized Haagy's work, saying Haagy's results were not consistent with SRI's findings. The speaker expressed regret that a respected scientist such as Dr. Haagen-Smit could make such a serious mistake and said, "It's unfortunate that a good chemist like Haagen-Smit could be misled."[98]

"Well," said Beckman, "I could almost feel Haagy's blood pressure rise. He was furious. The validity of his work was being questioned!" As they left the room, Haagy muttered, "I'll show them who's right and who's wrong."[99]

People who knew Haagy well spoke of his high integrity, so his muttered vow to show who's right or wrong might be entirely his personal reaction to the speaker's criticism of the integrity of his research. But it is likely that Haagy also was very aware that accusations his research was flawed—even if they were factually untrue but allowed to stand unchallenged—could destroy his career and seriously sully Caltech's reputation as a university. Witness events in 2023 when accusations of flawed research by the president of Stanford University were confirmed. Investigators found no evidence that he himself had engaged in any fraud or falsification, no evidence that he manipulated data—or knew at the time that subordinates manipulated data. But investigators said he did not correct the scientific record when concerns about errors in at least five papers surfaced later. He resigned his Stanford presidency, saying the investigation identified areas where he should have done better.

So, accusations that Haagy's research was faulty were serious—a "big deal"—very important to both Haagy and Caltech.

"People said I was guessing," Haagy said. "They talked about my 'theory.' But I wasn't guessing. I had worked long and hard to find it, and I was right."[100]

For some small idea of how long and hard, and the complexity of what Haagy did, scan the following brief excerpt from Haagy's June 1952 article "Chemistry and Physiology of Los Angeles Smog":[101]

> *The enzyme peroxidase, in the presence of hydroperoxide groups, catalyzes the oxidation of phenols and amines to colored derivatives. The oxidation of gualac tincture produces a blue color which can be measured colorimetrically. No color appears when the enzyme is poisoned with small quantities of hydrogen cyanide. This inhibition effect distinguishes peroxides from oxidation agents such as nitrogen dioxide and ozone, which produce a color regardless of the presence of hydrogen cyanide.*
>
> *The peroxidase test was run on an hourly basis.*

In response to scientists claiming their smog-formation experiments did not replicate Haagy's results, Haagy and his research associate, Margaret Fox Brunelle, looked closely at what the other scientists were doing and then explained why their experiments failed to get Haagy's results.

Primarily, the other scientists were not thinking in microchemistry terms. They used too much of each ingredient. Haagy said, "They hadn't understood that they had to work with a concentration of a tenth to one or two parts per million. They were putting together a few thousand parts per million of hydrocarbons and nitrogen oxides and failing."[102]

Margaret Fox Brunelle recalled other mistakes. One scientist used test tube stoppers made of rubber instead of glass. Because rubber contains hydrocarbons, they contaminated the tests. Another used tubes contaminated by a previous experiment. "I told him, 'You didn't clean your equipment completely!'"[103]

Above: Margaret Fox Brunelle, the research associate who helped Haagy teach other scientists what they were doing wrong and how to successfully replicate Haagy's experiments. *Courtesy of Margaret Fox Brunelle.*

Left: Victorious Haagy. Fellow scientists cheered: "The brilliant theoretical analysis and sound experimental evidence by Haagen-Smit and associates on the Los Angeles smog has been confirmed." *Courtesy of the Archives, California Institute of Technology.*

That was uncharacteristically blunt for Margaret, who usually was patient and kind. At home, she would drop whatever she was doing when she saw one of her cats gazing out the front door. "Oh! You want to go out!" and she would open the door and then open it again twenty minutes later—dropping whatever she was doing—when she saw the cat back at the door and gazing into the house. "Oh! You want to come in!"

This also was when Haagy discovered that smog formed only when the ingredients mixed in tiny amounts and the amounts had to be just right—not too little or too much. That was why other scientists failed to find smog right next to refineries or auto exhausts. They had to conduct their experiments farther away to allow the raw ingredients time to drift away from their source, time to mix with air, time to become diluted and weaker. That explained why drivers on the freeway sometimes did not see dense smog on the roadway but could see a blanket of shimmering silver-blue smog to the right and left about fifty yards away from the road.

After Haagy and Margaret coached them, fellow scientists' criticism finally changed to cheers: "The brilliant theoretical analysis and sound experimental evidence by Haagen-Smit and associates on the Los Angeles smog has been confirmed."[104]

Haagy discovered people first noticed the strong silver-blue smog in 1943 because petroleum refineries began to use a technology breakthrough to make *catalytically* cracked gasoline that produced twenty times more smog per gallon than did the old-style *thermally* cracked "straight-run" gasoline.[105]

Crude petroleum is a mixture of multiple hydrocarbon compounds that can be separated by distillation into "fractions" that have different boiling points. The fraction called gasoline originally was distilled by heating crude oil longer than required to make kerosene. In 1913, chemical engineers William Burton and Robert Humphreys of Standard Oil patented a refining method known as thermal cracking that applied both heat and pressure to increase how much gasoline could be distilled from petroleum. The procedure broke down—or "cracked"—heavy petroleum molecules. Such gasoline has been known as "straight-run" gasoline.

Catalytic cracking has been known since 1877 and was attempted commercially in 1915 by Almer M. McAfee using a batch process, but the cost proved prohibitive.

In 1922, French engineer Eugene Jules Houdry began working with E.A. Prudhomme, a French pharmacist who was producing exceptional gasoline using a catalytic process. Over the next few years, they developed an

Haagy said releasing gasoline and nitrogen oxides into the sunny greenhouse created a blue haze completely obscuring vision over a distance of only eight feet. *The LIFE Picture Collection via Shutterstock.*

improved production process, but in 1929, their demonstration plant proved too expensive to operate.

In 1930, Houdry was invited to come to the United States, and in 1933, he put a small Houdry catalytic cracking unit into operation. But again, the process was too expensive to survive the Great Depression.

Over the next few years, the Houdry process saw more improvements. About 1936, Sun Oil built a Houdry catalytic cracking plant. After running it secretly for about eighteen months, Sun Oil announced its success. The Houdry process doubled the amount of gasoline produced from each barrel of crude oil. Almost 50 percent of the product from Sun

Oil's refinery was gasoline, compared to 25 percent from the conventional, straight-run thermal process, and the gasoline also had a higher octane rating and burned more efficiently. The Houdry process gasoline became known as "cracked" gasoline.

Details of the successful Houdry commercial process shared at a 1938 meeting of the American Petroleum Institute reportedly "astounded" the industry. Eight more Houdry units were already under construction. By 1940, fourteen Houdry units were in operation. Refineries began using catalytic cracking as a batch process and then in 1942 implemented an improved continuous process.

With hindsight, the placement of two news stories in the *Los Angeles Times* on October 19, 1943, is interesting. The major story's headline seemed to deplore smog, saying, "City Files Suit to Halt Fumes from Aliso St. Rubber Plant." The story in the adjacent column seemed to cheer catalytic cracking. Its headline read, "Gigantic Plant Nearly Ready." It reported that Union Oil Company was about to open a new catalytic cracking refinery.

Haagy used a greenhouse to educate people about smog. "A student of mine and I got some greenhouse frames at a junkyard, put them together, and put in Plexiglas. It cost the tremendous sum of $350."[106] Releasing gasoline and nitrogen oxides into the sunny greenhouse created within a few minutes "a blue haze completely obscuring the vision over a distance of only eight feet."[107]

The greenhouse, Haagy said, "became a Mecca for politicians, industrialists, and friends. Fumigating them with a good dose of artificial smog was much more effective than any theoretical explanation I could have made."[108]

In the early 1950s, after his initial analysis of air for the Air Pollution Control District, Haagy returned to his Caltech laboratory to continue research into smell and taste. "I was finished with [air pollution]," he said. "This problem was man-made and I wasn't too interested in it. I have always been more interested in the problems that nature gives you. Why does a plant grow? Why does a fly have red eyes?"[109]

Chapter 8

FACTORY-IN-THE-SKY

What emerged from Haagy's research was a turning point for smog control. For many people it was revolutionary, a completely new idea: the concept of a factory-in-the-sky.

In the past, scientists and engineers like Dr. Uhl, Professor Tucker, Dr. McCabe and Dr. Weidlein had been looking for objectionable substances spewed into the air. Now, Haagy had discovered that normally unobjectionable substances spewed into the air reacted with each other and formed a new substance that was objectionable—smog. The seemingly fresh air all around us could now be described as actually like the frothing cauldron of Macbeth's three witches, a cauldron combining normally harmless substances into a new creation: harmful silvery-blue Los Angeles smog.

The recipe for making Los Angeles' smog has three ingredients: hydrocarbons, nitrogen oxides and sunlight. This recipe gave a formal name, a technical name, to Los Angeles' smog: *photochemical* smog. (A fourth factor played an important role: the inversion layer that held smoggy air over the city day after day.)

Let's be clear: sunlight is critical. Los Angeles' photochemical smog may be a unique form of air pollution. It exists only in the daytime, never at night. When the sun goes down, the chemical raw ingredients remain in the air but do not form photochemical smog.

To Haagy, the source of hydrocarbons seemed obvious but culturally delicate. "For a long time," he said, "we avoided the word 'gasoline' and called it hydrocarbons."[110] Smog was being created by automobiles, which were a key element of life in Los Angeles.

The historic 1954 factory-in-the-sky graphic explanation showing how harmless chemicals spewed into the air mix and react in the presence of sunlight to create obnoxious photochemical smog. *Image by Los Angeles County Air Pollution Control District.*

Ultimately, Haagy explained Los Angeles' photochemical smog as follows.

Morning commuter traffic creates a cloud containing the raw ingredients to make photochemical smog: hydrocarbons and oxides of nitrogen. Both are spewed from auto exhaust pipes. That cloud drifts from downtown toward the suburbs. Brilliant sunlight shines on the mix and changes it into photochemical smog. By evening, the east wind blows the photochemical smog cloud back through downtown Los Angeles. It stops west of the city along with pollutants created by evening commuter traffic (not enough sunlight then to convert them into photochemical smog).

For days, the cloud moves back and forth through the Los Angeles air basin, so raw ingredients for photochemical smog are in the air all over town every morning, just waiting for the sun to rise and shine. The chemical reaction operates only in the presence of sunlight (or strong artificial light under laboratory conditions). Up comes the sun, and then it is instant photochemical smog everywhere at once.[111]

That's why health inspectors in July 1943 discovered eye irritation began simultaneously at points ten miles apart and both downwind and upwind of the butadiene plant.

But knowing the recipe nature uses to make photochemical smog and explaining how it is produced every day in Los Angeles' sky did not solve the city's smog problem. Los Angeles was still the smog capital of the world.

Thermometers measure temperature in degrees. Hygrometers measure atmospheric humidity in percent. Barometers measure air pressure in inches of mercury's weight. Anemometers measure wind speed in miles per hour. But no instrument measured photochemical smog.

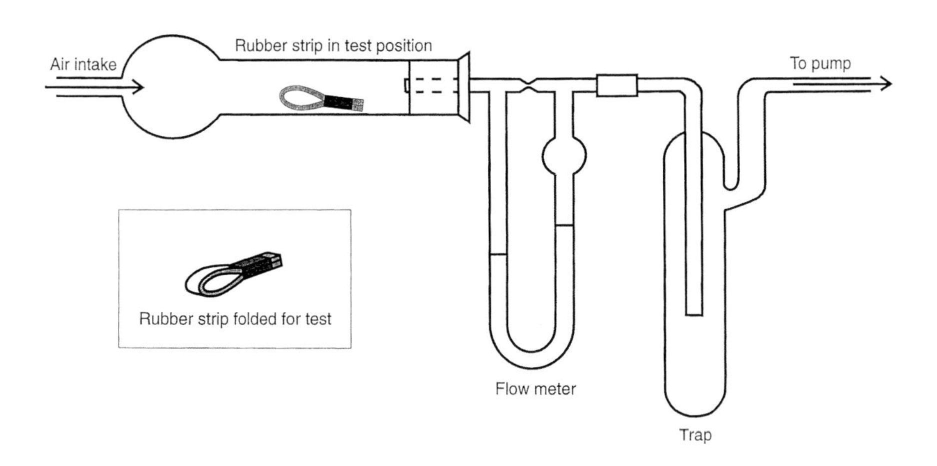

Above: To test photochemical smog's oxidizing strength, Haagy and Dr. Bradley pumped air over a folded strip of rubber and measured the time until cracks appeared. *Author's collection.*

Left: Ozone cracks rubber when the rubber is under tension. The magnifying glass shows a test rubber strip folded to create standardized tension for smog test. *Courtesy of University of Southern California, on behalf of the USC Libraries Special Collections.*

Haagy earned recognition for developing a rubber cracking test that measured photochemical smog by the effects it caused on rubber strips under tension. Haagy said Dr. Charles E. Bradley, his research associate, actually developed the technique. It passed measured streams of air across rubber strips stressed by being bent. Cracks in the rubber were caused by ozone manufactured in the air during the photochemical smog process. On a "non-smog day," the rubber showed cracks in half an hour to one hour. When smog was maximum strong, cracks were visible after seven minutes.[112]

The rubber cracking test caught the eye of the U.S. Armed Forces, which asked Haagy to study deterioration of rubber exposed to atmospheric influences over a period of years. An important practical application would be establishing guidelines for outside storage of rubber tires and other rubber parts to minimize the damage air—including smoggy air—could do to them.

Generally, observation of rubber goods marketed in the Los Angeles area showed more severe cracking than in any other part of the United States. To develop an inexpensive, reliable test indicating the cumulative effect of exposure to outside air, Haagy worked the problem with his research associate, Margaret Fox Brunelle, and his son, Jan W. Haagen-Smit, designing a test that exposed standardized rubber test strips to outside air for a period of one week and then measured the sum of the depth of all the cracks.[113]

Haagy's research showed the armed forces that one year of exposure to Los Angeles' air was equivalent to three to five years of exposure in other cities (Detroit, Yuma, Philadelphia, Dayton, San Antonio, Fort Greeley and Fort Churchill).

Firestone Tire and Rubber Company chemist Jay Elwood reported that his company, a maker of tires for automobiles, followed up on Haagy's research by adding a chemical to its rubber that protects tires from smog damage and cannot be washed, steamed or rubbed off.[114]

But knowing photochemical smog damaged tires and improving the tire rubber recipe to protect tires from smog damage did not solve the city's smog problem. Los Angeles was still the smog capital of the world.

For more than five years, from January 31, 1947, to March 14, 1952, the *Los Angeles Times* published a daily forecast of visibility at the Los Angeles Civic Center. In a nutshell, it forecast how visibility would be described using common terms like poor, fair, good or very good at various times of day, and it published the previous day's visibility ranges as compiled by the U.S. Weather Bureau. An illustrative example:

Forecast
Sunrise to 8 a.m.......Poor
8 a.m. to 11 a.m.......Poor
11 a.m. to 2 p.m.......Fair
2 p.m. to sundown.......Fair
Yesterday's Range
6 a.m.......¾ mile
9 a.m.......1 mile
Noon.......2 miles
(Haze, 6 a.m. to noon)
3 p.m.......15 miles
6 p.m.......15 miles

Later, Henry Dreyfuss also was interested in forecasting smog. He asked Haagy to help design an international weather symbol for photochemical smog.

Dreyfuss built his career around designing products that would appeal to customers. He began in about 1921 by designing Broadway theater sets and through the years designed products many people used: Bell telephones, the Princess telephone, a General Electric home refrigerator, Polaroid cameras, home thermostats, Hoover vacuum cleaners, Westclox's Big Ben and other alarm clocks, a Singer sewing machine, a Royal typewriter, a John Deere tractor, the J-3 Hudson railroad locomotive and towers to carry high-voltage electric power lines.

While examining standard meteorology symbols, Dreyfuss noticed none for smog, and the U.S. National Weather Service confirmed no symbol for smog existed. Dreyfuss found it "absurd to have no graphic symbol for so prevalent and dangerous a nuisance" and sat down with Haagy—whom he referred to as "Mr. Smog-control"—over a lunch table to develop a graphic symbol: a valley shaded black for heavy smog, medium gray for medium smog, light gray for light smog and clear for no smog, with a circle over the valley to symbolize the essential effect of the sun (no sun, no smog; photochemical smog does not form between sundown and sunrise). Haagy displayed the set of symbols at least once, and Dreyfuss printed them in a book, but there is no evidence they were ever adopted by any weather forecasting agency.

Forecasting visibility, reporting the previous day's visibility ranges and having a graphic symbol to indicate heavy, medium, light or no photochemical smog might be useful but did not solve the city's smog problem. Los Angeles was still the smog capital of the world.

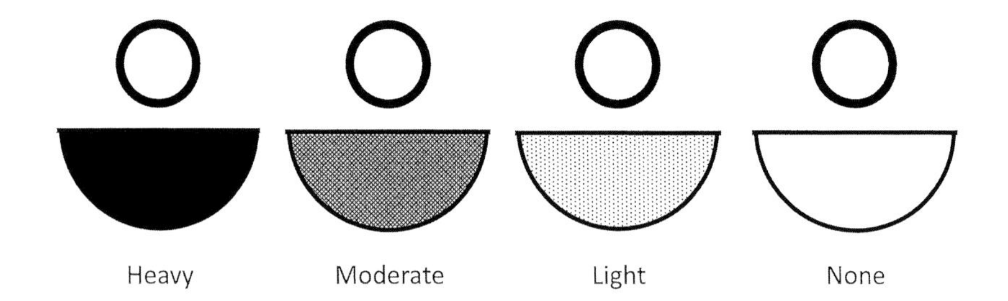

Designer Henry Dreyfuss and Haagy created a weather symbol for smog: a circle for the sun above a valley shaded to show the smog's strength. *Author's collection.*

The Air Pollution Control District undertook to reduce or eliminate evaporation of gasoline at petroleum refineries. The district had clear enforcement power there, and control devices were well known and simple. Still, longtime Los Angeles County supervisor John Anson Ford expressed frustration. "The refineries submitted sometimes reluctantly to regulations promulgated by the district's engineers and enforcement officers."[115]

Dr. Paul Mader, an Air Pollution Control District chemist, developed a method to accurately measure gasoline vapor in the air. It showed a single skimming pond lost forty tons of gasoline to evaporation daily. Annually, all Los Angeles County refineries lost $30 million worth of gasoline to evaporation each year—this when gasoline retailed for about $0.28 per gallon, including taxes. Over time, refineries improved the refining process, reduced venting of waste gases into the air, installed floating covers on petroleum tanks and captured and condensed gases that otherwise would have evaporated into the air.

That gave the companies two benefits. First, on the public relations side, they won a good reputation as citizens meeting their duty to not pollute the air. Second, on the profit side, preventing evaporation put $30 million of extra money in their pockets year after year.

Nevertheless, while floating lids and capturing gasoline vapors at refineries seemed important and useful, they did not solve the city's smog problem. Los Angeles was still the smog capital of the world.

Beginning in the 1940s, scientists saw damage to trees but could not identify the cause; they called it "X disease." In 1944, farmers in the Los Angeles area reported damage to leafy vegetables. In the spring of 1949,

Los Angeles Daily News photo of a smoggy street in 1953 as people realized the smog continued to get worse. *UCLA Special Collections CC by 4.0.*

crops showed damage variously described as a silvery, metallic appearance; a metallic glaze; and, sometimes, brown, dried spots on the leaves or irregular bleaching. The ailment affected pinto beans, tomatoes, garden and sugar beets, Swiss chard, eggplant, head lettuce, endive, spinach, romaine, alfalfa, tobacco, petunias and oats. Farmers reported losing whole fields of beets, spinach, endive and romaine.

Dr. McCabe arranged for scientific investigation by Caltech and the Riverside Citrus Experiment station. Haagy's Caltech colleague Frits Went joined Riverside plant physiologists John Middleton, James Kendrick Jr., Ellis Darley and other scientists in research conducted primarily at Caltech's then brand-new Earhart Plant Research Laboratory and also at fumigation cubes at the University of Riverside and at exposure boxes placed in five towns in the Los Angeles air basin.

A visitor to the Earhart Laboratory would be surprised to encounter clean-room procedures designed to prevent contaminants from entering testing laboratories. Everyone entering needed to strip down, wash and put on sterile clothing: white pajama pants, a white slip-on shirt and a white cap. Lab director Went said a single contaminant could cause up to $10,000 in damage.

Dr. Went called the lab a phytotron, essentially a weather factory. Switches and buttons on a control panel created custom climates in thirteen laboratory rooms equipped with lights that could simulate sunshine, various types of water—cold, hot, tap, de-ionized—nutrients and variable air pressures.

Initially, Dr. Went systematically tested fifty-five contaminants known to be in the air of Los Angeles but found nothing that would damage plants at concentrations actually found in the atmosphere. He mentioned the problem to Haagy, who suggested testing photochemical smog.

Haagy joined the project. A test in the Caltech plant laboratory exposed a spinach plant to smog, then alfalfa, sugar beets, oats and endive. All showed damage like field-grown plants suffered.[116]

By 1954, smog damage affected forty-four commercial crops. Researchers saw brown patches on head lettuce leaves, brown spots and bronzing or silvering on romaine and spinach and brown ends on alfalfa leaves. Beets showed silvering, bleaching and pitting. Celery and mustard became less green. Western yellow pine trees shed needles and died prematurely. Commercial farmers stopped planting spinach, celery, lettuce, tomatoes, string beans and cucumbers in Los Angeles and then in surrounding counties as well.

But evidence showing smog damaged plants did not solve the city's smog problem. Los Angeles was still the smog capital of the world.

In the mid-1940s, some authorities asserted that Los Angeles' smog was obnoxious but not a serious health hazard. Scientists tested that assertion using three kinds of studies: epidemiological comparison of pollution levels with health statistics; experiments exposing monkeys, dogs, rats and other animals to controlled amounts of air pollution in laboratories; and experiments exposing healthy human volunteers to levels of pollution involving no risks known to be serious. Those studies indicated air pollution could be linked to human illness and sometimes to premature death. Both

Dr. Frits W. Went, director of Caltech's Earhart Plant Research Laboratory, at the control panel able to adjust experiment conditions: sunshine, water, nutrients, air pressure. *Used with permission of the William R. "Bill" Watson Estate.*

men and women, of all ages, could be affected, but risks were higher for the very old, the very young and people already sick with chronic ailments.

Air pollution aggravated respiratory system problems—nose, sinus, throat, bronchial tubes and lungs—and was linked to diseases of the heart and blood vessels. Symptoms of pollution exposure included coughing, wheezing, phlegm, shortness of breath, chest discomfort or pain, nausea,

Above: Haagy, Paul Mader and Gordon Larson checking the effect of artificial smog on plants in a test chamber. By 1954, smog had damaged forty-four commercial crops. Los Angeles Times *Photographic Archive, UCLA Library Special Collections.*

Opposite: Haagy checks a flue gas re-circulating system at El Segundo electricity generation plant to help the Edison company clean its smokestack emissions of pollutants that create smog. *Image by Los Angeles County Air Pollution Control District.*

headaches, eye irritation and dizziness. Long-term effects for residents of high pollution areas included decreases in lung function when compared to residents of low pollution areas.

But studies showing smog harmed animals and humans did not solve the city's smog problem. Los Angeles was still the smog capital of the world.

The local electric company, Southern California Edison, persuaded Haagy to leave his laboratory in 1956 to work on smog control again. He took a two-year leave of absence from Caltech to help. Edison had just built a brand-new oil-fueled steam station to generate electric power at El Segundo, one of the towns in the Los Angeles air basin. The plant met standards approved by the Air Pollution Control District and was a prototype for other stations Edison planned to build.

However, when the new plant fired up, its smokestack plume looked darker than allowed on the Ringelmann chart. So, the Air Pollution Control District would not issue a permit for it to operate. Edison was stuck with a multimillion-dollar albatross. Edison appealed to an arbitration panel, which allowed Edison to operate the El Segundo plant on condition that an investigation be undertaken to determine if and how the opacity of the smokestack plume could be reduced.

Edison sought to do more than just reduce plume opacity. It wanted to eliminate dust, sulfur oxides and nitrogen oxides—any emission that might ever offend pollution inspectors and give them reason to shut down the station in the future.

Haagy earned credit for "a signal role" in developing a technique to reduce nitrogen oxide emissions from power generation stations by using two-stage combustion.[117] He helped Edison design equipment to eliminate offensive emissions, which they tested successfully at two sites, El Segundo and Etiwanda. The million-dollar investment set a new standard of combustion efficiency and cleanliness for the electric power industry. Haagy pointed out that Edison's solution not only solved their local problem but also could be applied nationwide.

While eliminating offensive emissions at power generation stations seemed important and useful, it did not solve the city's smog problem. Los Angeles was still the smog capital of the world.

It was clear to everyone in Los Angeles that the huge unsolved problem that caused Los Angeles' smog to continue getting worse and worse, the overwhelmingly major cause of photochemical smog, was automobiles. Automakers were the only ones who could fix that problem, but no one seemed able to get the automakers' attention.

Chapter 9

THE BIGGEST CAUSE OF LOS ANGELES' SMOG

Los Angeles was losing its war on smog.

People were growing more impatient and more upset.

Smog would not go away.

Los Angeles County supervisor Kenneth Hahn stated the problem plainly: automakers were not cooperating. They were failing their duty "to the American public and to the health of the community....[W]e have urged the industry that causes the [photochemical smog] problem to cure it. The industry, in my judgment, did not care one bit about the health of the community as far as air pollution for many years."[118]

Hahn began writing to major auto manufacturers—a letter-writing campaign that was to last fifteen years. Auto executives wrote back to Hahn, saying they did not feel autos caused Los Angeles' smog. "The Ford engineering staff, although mindful that automobile engines produce exhaust gases, feels these waste vapors are dissipated in the atmosphere quickly and do not present an air-pollution problem."

General Motors wrote, "[S]tudies made by various investigators in Los Angeles have indicated that certain hydrocarbon effluents from automobile exhaust gases may be a contributing factor to the smog and particularly to the eye-smarting components of the smog. As far as we are aware, Los Angeles is the only community having this particular complaint...indicating that perhaps some other factors other than automobile exhaust gases may be contributing to this problem."[119]

Left: County supervisor Kenneth Hahn told Detroit automakers they were failing in their duty to stop cars from spewing pollutants that create Los Angeles' smog. Herald Examiner *Collection/Los Angeles Public Library.*

Below: Auto engineers arrived five years after Haagy discovered Los Angeles' smog recipe to see what automakers could do "to help cure this local problem." Los Angeles Times *Photographic Archive, UCLA Library Special Collections.*

Opposite: The doll was masked too when Agatha Acker, three, joined her mother and sisters in a Pasadena, California park for an antismog demonstration. *Bettmann via Getty Images.*

In about January 1954, the Detroit automakers created a multi-company team to explore motor vehicle contributions to smog. Chrysler vice president James C. Zeder, director of engineering and research, said people might be surprised that the companies were working together in the battle against smog because the auto industry has a reputation for being fiercely competitive. Ordinarily, Zeder said, competition in research, engineering, production and sales is the best way, but sometimes energetically sharing knowledge and helping each other with the workload solves a problem quicker.

Finally, on January 25, 1954—five years after Haagy discovered the recipe for Los Angeles' photochemical smog—a group of auto engineers arrived in Los Angeles to make the auto industry's "first 'on the spot' survey to see what it can do to help cure this local problem."[120]

But the engineers brought with them the usual definition for smog: "a combination of smoke and fog." And when they arrived by train, they saw near the Los Angeles railroad station "a huge incinerator belching out smoke. Surely this was the source of the 'smog.'"[121]

In the end, the automotive engineers' reaction to their visit was that the next three weeks were "a revelation."

The engineers encountered a haze that smelled like ozone, and on a couple of days there was severe smog. Haagy, "a painstaking scientist, and a very pleasant gentleman," explained how nitrogen oxides, hydrocarbons from gasoline and sunshine combine to create Los Angeles' photochemical smog.[122]

The engineers were impressed to discover that smog was a "high level deal" in Los Angeles. People personally involved in their discussions of smog were chairmen of company boards, presidents, vice presidents and top politicians. That was a surprise.

Near the end of their visit, the auto engineers built a demonstration based on the definition of smog they had arrived with. They burned a gasoline torch that had no visible smoke to simulate an automobile. And they burned a smoke pot containing gasoline, oil, vegetation clippings and hamburger

Highland Park Optimists ironically wore gas masks at a 1954 meeting to protest that no one was doing anything to fix Los Angeles' smog problem. *UCLA Special Collections CC by 4.0.*

that emitted a dark smoke to simulate the hundreds of incinerators then burning in Los Angeles.

Had they learned nothing during their visit?

The audience reacted vehemently. Speaker after speaker got up to forcefully point out that smog posed a tremendous health menace; the auto industry was naughty, acting irresponsibly and indefensibly; and auto engineers should quit worrying about pollution sources other than cars and fix the cars.

When they returned to Detroit, the engineers confirmed that devices to stop automobiles from producing Los Angeles' photochemical smog had not yet been invented. Worse, the automakers did not have the microchemistry instruments and methods they needed to search for the microscopic photochemical smog ingredients.

General Motors scientific director John Campbell described his view of the auto industry's problem. Very little gasoline fails to burn in the engine. Only a few tenths of 1 percent go unburned and appear in exhaust gas. And the proportions of exhaust gas ingredients change constantly as engine load flexes to meet conditions encountered in normal driving. The objectionable substances they were looking for were present as only a few parts per million—microchemistry proportions.[123]

After several of its motorcycle couriers became ill in September 1955, Rapid Blue Print Company bought gas masks to protect them from smog. *Bettmann via Getty Images.*

An antismog demonstration in 1956 outside the Federal Building in Los Angeles to protest that no one was doing anything to fix Los Angeles' smog problem. Herald Examiner *Collection/Los Angeles Public Library.*

In fairness, the automakers did invest big money to develop new research equipment. But Haagy discovered their equipment sometimes was "too good" and led them to wrong conclusions. Haagy sometimes got better results using the simple, cheap equipment he could afford.

For example, their expensive equipment seemed to show General Motors that increasing engines' nitrogen oxide emissions would decrease smog. But Haagy knew that approach would not work. People dislike high concentrations of nitrogen oxide. "My simple fumigation room with its manual adjustment of flowmeters made me a part of the experiment while GM's elaborate equipment [operated entirely from outside the lab] did not allow the experimenter to [experience conditions inside and] enjoy their suggestion."[124]

Chrysler research director James Zeder pointed to a major challenge automakers face: new devices—certainly new smog control devices—require extensive testing and improvement to ensure they are practical. They must operate for hundreds of thousands of miles, in good and bad weather, in broiling heat and bitter cold, when shaken over rough roads, at sea level and in the thin air of mountain peaks. Bitter experience taught automakers such testing cannot be bypassed. Even then, "some drivers will manage to get into situations we did not foresee."[125]

Automakers wanted time; politicians and Angelenos wanted immediate solutions. Supervisor Hahn took a leading role in organizing legal action to force antismog improvements by automakers, culminating in an antitrust case in federal court that alleged automakers secretly agreed not to produce a car equipped with new pollution control devices before mutually agreed-upon dates, lied when they said it was technologically impossible to introduce new exhaust controls and agreed among themselves to delay such controls. But the case never went to trial. It was settled out of court with a consent decree in which automakers, while admitting no guilt, promised not to conspire to obstruct development of emission controls and agreed to make air pollution control equipment available to all companies royalty-free.[126]

Los Angeles County supervisor John Anson Ford said Los Angeles officially asked automakers to solve the exhaust emission problem. The auto companies responded that they were working hard on the difficult problem but so far had not come up with a muffler or exhaust that met the need and three or four more years might be required before new cars could be equipped with such devices even after they were perfected.

When those "three or four" years passed with no results, the Los Angeles County supervisors decided to replace persuasion with legal prohibition.

Stamp Out Smog demonstration outside a board of supervisors meeting in 1969 protesting that no one was doing anything to fix Los Angeles' smog problem. Herald Examiner *Collection/Los Angeles Public Library.*

Southern California officials met with Governor of California Edmund G. "Pat" Brown. Governor Brown sponsored a state air pollution law that created a statewide Motor Vehicle Pollution Control (MVPC) Board that could *require* antismog devices on cars that were to be sold in California.

Los Angeles County supervisor John Anson Ford tried to persuade Detroit automakers to solve the exhaust emission problem. Years passed, and then legal prohibition replaced persuasion. Los Angeles Times *Photographic Archive, UCLA Library Special Collections.*

The MVPC Board—the first state agency in the United States to control motor vehicle emissions—set standards and then tested cars. If a model polluted too much, the board could order the manufacturer to recall and repair its cars. The board could even—its maximum penalty—order the automaker to stop selling cars in California.

In those years, Detroit manufacturers dominated the U.S. auto market, and auto companies had seen Los Angeles as a one-and-only small voice raising a complaint about smog. But the reality was that California was a major market. Californians bought about one of every ten cars Detroit produced. Now the entire state of California stood with Los Angeles, and eventually many of the controls California originated would become the foundation for a U.S. federal control program that began in the 1960s.

California asked Haagy to serve as a member of its Motor Vehicle Pollution Control Board. Haagy felt uneasy. "I felt I was competent in chemistry, but not in government. However, the job had to be done. I never walk away from anything."[127]

Other people also were uneasy about Haagy's untested government skills. Clean air environmentalist Gladys Meade said "he was second choice. [The first-choice person] had a conflict of interest and couldn't serve."

But Meade said people were delightfully surprised at what Haagy could do. "He was very politically aware." She felt Haagy deserved great credit for getting reluctant automakers to invent and install pollution controls on their cars.[128]

Haagy's method was to go straight to the top. "I have found that it does no good to go to anyone unless he is in charge. All a company president has to do to get something done is send out a little memo."[129]

Engineers discovered unburned gasoline routinely escaping from each and every auto—most from the exhaust pipe (65 percent), some from the

Tomorrow, and Tomorrow and Tomorrow…A theme editorial artist Herblock borrowed from Shakespeare's *Macbeth* to highlight little to no progress. *A 1976 Herblock Cartoon, © The Herb Block Foundation.*

Governor Brown with President Jack Kennedy. Pat Brown signed the law creating a Motor Vehicle Pollution Control Board that would require antismog devices on cars. *Abbie Rowe, White House Photographs, John F. Kennedy Presidential Library and Museum, Boston.*

crankcase (20 percent) and some from the carburetor and fuel tank (15 percent). Every car in motion puts about 1 pound of hydrocarbons into the air each day. For the 3.5 million autos in Los Angeles County, the total pollutants escaping into the air each day were 1,625 tons of hydrocarbons and 485 tons of nitrogen oxides.[130] Moreover, fumes were escaping at retail gas stations when drivers refueled cars at the gas pumps and when tanker drivers refilled station storage tanks. Those problems were fixed by:

- Sealing autos' gasoline tanks (which required adding chambers to the tank to allow gasoline to expand on warm days and vapor-liquid separators to return liquid to the tank and pass vapors through a valve to an activated carbon canister purged by a vacuum created by engine operation that carried vapors into engine combustion chambers to be burned).
- Installing a closed crankcase ventilation system to capture gasoline vapors leaking between piston rings and cylinder walls and feed them back into the combustion chamber.
- Replacing carburetors with fuel injection systems that improved delivery of gasoline into cylinders.
- Redesigning combustion chambers to increase the percentage of fuel successfully burned and minimize unburned gasoline sent into the exhaust.
- Booting. "Empty" auto gasoline tanks actually are full of vapors. Gasoline delivered into the tank forces those vapors out. Boots

Los Angeles' smog capital of the world reputation led Shriners visiting in 1960 to form a "smog detail" wearing gas masks. Herald Examiner *Collection/Los Angeles Public Library.*

> installed on filling station nozzles capture those vapors during fill-ups and pipe the vapors into a tank at the station instead of allowing them to spew into the air. Similarly, trucks delivering gasoline to filling stations got a second hose because an "empty" tank at the filling station actually was full of vapors from the gasoline that used to be in it. In the past, filling the station's tank forced those vapors into the atmosphere. Now the vapors were captured by hose two and captured in the delivery truck.

By 1972, 90 to 95 percent of the cars on California roads had antismog devices of some sort, and smog had decreased for five or six years. Haagy was proud of that result. "[F]orgetting the time it took, it is a rather impressive record, considering that we have not wrecked the automobile industry and we still have acceptable automobiles."[131]

However, auto exhausts were still spewing smog ingredients into the atmosphere. Was there a practical, cleaner alternative to the internal combustion engine?

A forty-foot urban bus running on an external combustion steam engine was tested in 1972. *Courtesy of L.A. Metro.*

The federal government and the California State Assembly sponsored two projects to explore alternative motor vehicles that would create less smog. In 1968, they first designed and installed steam engines into standard forty-foot urban buses. Testing showed exhaust emissions and exterior noise levels were both desirably low, but interior noise and fuel consumption were both undesirably high.

The buses were followed by a second project that designed steam engines for automobiles. Contractor Aerojet Liquid Rocket Company installed its engine in a 1973 Chevrolet Vega hatchback coupe. Steam Power Systems, Inc., built its own car.

In May 1974, the two steam cars were demonstrated to a few people outside the capitol building in Sacramento, California. News reports said one car refused to budge, apparently due to transmission problems. The other took five minutes to get started. When running, they sounded like vacuum cleaners. The *Los Angeles Times* described the demonstration as "trouble plagued."

California Assembly speaker Bob Moretti characterized the auto project, which cost $2.75 million in public funds, as the "start of a movement" to get the Big Three (General Motors, Ford and Chrysler) off their backsides to do something about developing clean engines.[132] The California steam cars reportedly got nine miles per gallon in city driving and thirteen miles per gallon on the highway.

Automakers reported their own experiments with steam power showed less air pollution but problems that were enormous: heavy weight, risk of steam engine water freezing in cold weather and low miles per gallon.

An electric-powered "Volts-Wagon" truck driving through smog in Los Angeles in 1967 during testing by the Los Angeles Department of Power and Water. *UCLA Special Collections CC by 4.0.*

Lead-acid batteries in the trunk of an electric car tested in 1967 by the Los Angeles Department of Power and Water. Herald Examiner *Collection/Los Angeles Public Library.*

Experiments with other alternatives to gasoline and its polluted exhaust included vehicles powered by electricity. For a short time, the Los Angeles Department of Power and Water tested a Volts-Wagon truck and an Electrauto car, each powered by conventional lead-acid batteries.

People sometimes criticized Haagy's work on government air pollution boards, saying he was too weak or too tough. But he said he never let the shouters influence him one way or the other. "They got awfully excited, but I just stayed quiet. It's better not to talk back. You always lose."[133]

Haagy may have been politically aware, as described by Gladys Meade, but he expressed discomfort with the politics he admitted were a practical requirement to get the legislature to act. "It is very difficult for a scientist to agree with some of the ways this has been done." But he conceded the end may justify the means: "The unreasonable approach has success and the reasonable one doesn't get anywhere."[134]

Dr. Arie Haagen-Smit, the man who was not interested in studying air pollution because it was a problem made by people's activities rather than nature's, emerged as the guru of Los Angeles' smog, the leading expert, and the world would not let him leave that role to return to his experiments into nature—into smell and taste—that could lead to a Nobel Prize. That changed his life. Haagy said it forced him to trade the test tube for the telephone. "I suppose, it was not the intention of the fly to catch the flypaper either, if you know what I mean."[135]

In 1967, California changed the Motor Vehicle Pollution Control Board into an Air Resources Board, with Haagy as chair. In July 1972, the state reduced the Air Resources Board to five members, all experts in air pollution control. Haagy wanted to retire from the state job but was prevailed upon to continue, still as chair of the board. "Let's face it," said one board member, "his prestige alone was holding the whole thing together."

Haagy said he agreed to stay because "the public has a right to clean air. Perhaps I can help."[136]

The following year, 1973, Haagy did retire, saying the job had entered a new phase and needed a different type of person, someone able to knock heads together. Knocking heads was not Haagy's strength. His nature favored listening patiently to people, reasoning with them, being a quiet persuader. One associate characterized him as "the foxy grandpa." Haagy once was criticized for allowing someone to speak for forty-five minutes without saying anything new at an Air Resources Board meeting. Haagy's explanation: "He's a human being. He'd feel badly if no one listened."[137]

Thirty years after smog first struck, it still conceals the tallest buildings in downtown Los Angeles on October 21, 1973. *UCLA Special Collections CC by 4.0.*

Colleague James Bonner praised Haagy for being the very rare example of a true scientist who takes a problem and solves it both scientifically and also societally. Haagy told Bonner that he felt he had covered the field of smog research and control pretty well—the science, the legal, the political—but he believed he might have done more in city planning.

Bonner cheered. "It was one man against an establishment that at first insisted that petroleum and automobiles could not possibly be the source of smog—but as we all know, Haagy won and won totally."[138]

But Los Angeles still was losing its war on smog. Angelenos still were impatient and angry. Smog still would not go away. Los Angeles still was the smog capital of the world.

Chapter 10

SOLVING THE CATALYST PROBLEM

A surefire way to eliminate smog-causing emissions from auto exhausts, a solution obvious to most everyone but acceptable to hardly anyone, was to eliminate autos. But an aide to California governor Pat Brown said, "We're not about to abolish automobiles in California."[139] That idea was ludicrous, preposterous; a better idea must be found.

Auto engineers believed one device, among all their antismog inventions, promised the greatest success in eliminating photochemical smog. It would be the "gold standard" if they could get it working. But despite their best efforts, the device remained so impractical that no one could use it.

In March 1953, Haagy saw a demonstration of the Oxy-Catalyst Muffler. It was built by Dr. Eugene Houdry, the same man who made practical the catalytic cracking of petroleum that (unknowingly and unintentionally at the time) generated twenty times more photochemical smog than straight-run thermal cracking did. Houdry's device was demonstrated in Los Angeles and Detroit. It successfully removed hydrocarbons from auto exhaust, but the muffler was impractical because lead reacted chemically with the catalyst, quickly destroying it.

The recipe for gasoline used throughout the United States for three decades included lead.[140] For years, engineers searched for an affordable and effective alternative catalyst that would not be destroyed by lead, but they found nothing.[141]

An alternative to the catalytic "muffler" was tested: an "afterburner." It mixed pollution-filled exhaust gases with fresh air and set them on fire to

Los Angeles Times editorial artist Bruce Russell portrayed the ever-lingering Los Angeles smog problem as Public Menace No. 1 on January 8, 1961. *Used with permission of the Bruce Russell Estate.*

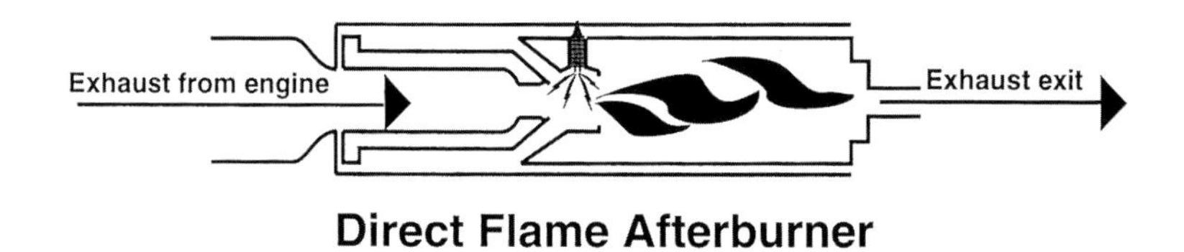

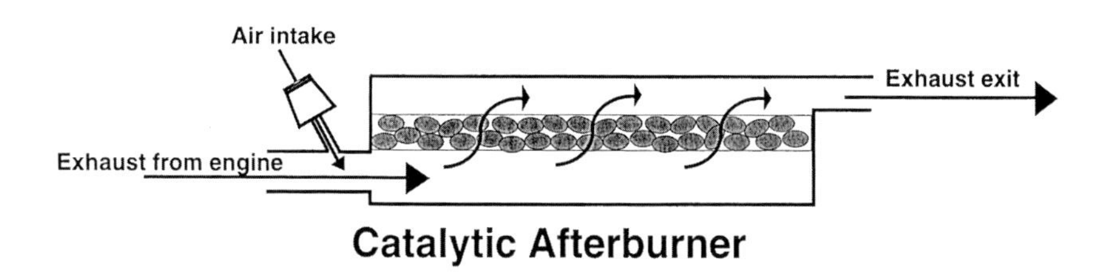

Two "muffler" ideas to eliminate hydrocarbons from auto exhaust: one would use fire to burn them, the other a catalyst—without flames—to destroy them. *Author's collection.*

consume unburned hydrocarbons. This afterburner device created a hot fire located under the car, something of a drawback.

Why was lead in gasoline? Up to the 1920s, auto engines suffered "knocking," which was gasoline igniting prematurely, before the cylinder was in position to use the explosive force. In 1921, General Motors auto engineers discovered that tetraethyl lead was a low-cost additive that prevented premature gasoline ignition. It would eliminate "knocking" and "pinging" and make auto engines run smoothly. Thereafter, the recipe for almost all gasoline sold in the United States included lead.

When you have eliminated all other options, when you have been unable to find any affordable and effective alternative catalyst that will not be destroyed by lead, then the "impossible" solution that remains, however difficult, must be the solution you use: eliminate lead from the recipe for gasoline.

A change to selling gasoline without lead would require many other changes. New auto engines would have to be designed to safely use unleaded gasoline without knocking. The many cars already on the road would still need leaded gasoline. That leaded gasoline would always have to be kept separated from unleaded gasoline. To do so would require new oil refinery equipment, new storage tanks, new trucks and new pipelines, and retail gas stations would need both new tanks and new pumps.

Los Angeles Times editorial artist Bruce Russell, using French artist Paul Émile Chabas' young woman bathing chastely in a lake, showed the impact smog has. *Used with permission of the Bruce Russell Estate.*

So a change to selling gasoline without lead would be huge, affecting the entire nation. The city of Los Angeles was too small to do it alone. The county of Los Angeles and the surrounding counties and the state of California, all combined, were too small to do it.

Finally, the U.S. Congress used its national power. It authorized the U.S. Environmental Protection Agency (EPA) to require sale of unleaded gasoline, and the EPA ordered the enormous change to begin nationwide on July 1, 1974.[142]

For most of the nation, the huge change was justified to eliminate childhood exposure to airborne lead from car exhausts that caused brain damage. For Los Angeles, the huge change also was justified as a vital pre-condition so automakers could be required to install effective catalytic "mufflers" on all new cars sold in California. In that way, the entire United States now stood with California and Los Angeles to fight Los Angeles' photochemical smog.

The EPA's July 1, 1974 rule did not stop the sale of leaded gasoline but did require all large retail gasoline stations throughout the United States to also sell unleaded gasoline. To prevent people from putting leaded gasoline into a catalyst-equipped car, the EPA required auto manufacturers to build a narrower tank inlet on cars that used unleaded gasoline and required gasoline stations to use correspondingly smaller nozzles on unleaded fuel pumps.

The automakers' deadline for building cars with catalytic "mufflers" was the 1975 model year. Those cars would begin to be sold in the second half of 1974. All 1975 model year cars sold in California had catalytic "mufflers," as required by that state. Moreover, 85 to 90 percent of all 1975 model year cars sold throughout the United States had them.[143] The devices today are known as catalytic converters.

In 1977, the Environmental Protection Agency sent people in unmarked vehicles to park inconspicuously near gas stations and observe 368 people in sixteen states filling the gas tanks of their cars. The EPA noted what gasoline pump they used (unleaded or leaded gasoline) and the license plate of their car. Later, they checked state auto registration records to verify whether the car was equipped with a catalytic muffler. They found about 88 percent of the people were properly buying unleaded gas for their car (priced a few pennies per gallon higher than leaded gas) and about 12 percent were violating the law by putting the less expensive leaded gas into their tank (which permanently destroyed the catalyst and voided the manufacturer's warranty on that catalyst). In every instance, the gas station nozzles were the correct large diameter for leaded gas and small diameter for unleaded.

What violators had done was modify the filler inlet on their car, removing the restrictor device that narrowed large pipes or expanding small pipes to accommodate the large diameter nozzle.[144]

An "air pollution emergency" occurred on June 27 and 28, 1974: Stage 3 smog alerts on two consecutive days in San Bernardino County along the eastern edge of the Los Angeles basin. California rules called for immediately and drastically reducing vehicle traffic by closing government offices, closing regional shopping centers and closing factories.

The *Los Angeles Times* reported some parks and playgrounds closed and some factories reduced output, but otherwise "almost nothing happened." The *Times* quoted Lin Koester, director of environmental health and air pollution control officer for Ventura County, as saying, "It's easy to say, 'Let's shut everybody down,' but it's a little harder to accomplish." William Simmons, executive officer of the California Air Resources Board, said, "The public attitude is, 'let somebody else do it.'" Haagy said air pollution alerts present "an impossible enforcement problem....It's an unrealistic thing."[145]

Those observations are consistent with a fundamental principle of organization development: people are complicated and very difficult to fix; process improvements (like equipping cars with catalytic converters) are comparatively simple, efficient and productive.

At an air pollution conference held twenty-five years after he discovered the recipe for Los Angeles' smog, Haagy commented that people expect clean air too fast and grossly underestimate the sacrifices they must make to get it. "The public wants clean air. Yes, they want clean air if they don't have to go to too much trouble."[146]

In 2023, the California Air Resources Board published a "history of California's pioneering efforts to reduce air pollutants" that praised Professor Arie Haagen-Smit, saying, "His research, highlighting the reaction of sunlight with automobile exhaust and industrial air pollution, became the foundation upon which today's air pollution regulations are based."[147]

The history noted that California's challenges include more people and more cars every year; nevertheless, "in the 1980s and '90s, California cars became the cleanest in the world, and California's fuel became the cleanest, too."

Los Angeles' terrible photochemical smog turned out not to be just a "local problem." Scientists found the same smog in many other cities, even in London, where "the very process of cleaning up [coal smoke smog from] the air...has meant more sunlight, and hence more photo-oxidation....London has begun to experience the smogs usually associated with Los Angeles."[148]

This *Hi and Lois* cartoon gave someone very young surprising insight and wisdom. Hi and Lois © *1994 Comicana Inc, Distributed by King Features Syndicate, Inc.*

By 1961, scientists had discovered photochemical smog affecting twenty-five other counties in California, twenty-four other states and six other nations—three continents in all.[149]

By October 1972, the Organization for Economic Cooperation and Development reported antismog rules imposed by more than seventy cities in western Europe. Included: in Vienna, a buses-only zone all day in the inner city; in London, the busy Oxford Street closed to all but buses and taxis; in Goteborg, Sweden, only buses and trolleys allowed to cross downtown.[150] At various places in the world, smog brings human activity to a halt—people are urged to shelter indoors until the air clears.

Generally, people are concerned about smog's effect on human health. But there are other concerns too. The government of Greece removed statues sculpted 2,500 years ago from sites outdoors on the Acropolis to indoor displays at the Acropolis museum. The reason? Erosion accelerated by air pollution. Greece received a warning from the United Nations Economic and Social Council (ECOSOC) that quick action was needed or "the monuments may be lost forever."[151]

Colleagues anticipated that Haagy's research into such things as the growth, aromas and flavors of plants like pineapple, garlic and wine grapes at Caltech would earn him a Nobel Prize, but he sidelined that research to discover the chemistry of smog and the sources of pollutants creating smog and thereafter to help lead efforts by the City of Los Angeles, the State of California and the United States to eliminate photochemical smog. There is no Nobel Prize for air pollution control.

President Richard M. Nixon invited Haagy to the White House on October 10, 1973, to be awarded the National Medal of Science for "unique contributions to the discovery of the chemical nature and source of smog, and for the successful efforts which he has carried through for smog abatement."

Haagy received the National Medal of Science from President Nixon for the discovery of the chemical nature and source of smog and successful smog abatement. *Oliver F. Atkins, White House Photo Office, National Archives.*

The *Los Angeles Times* published the last interview of Dr. Arie Haagen-Smit on Sunday, March 6, 1977. Reporter Al Martinez wrote that Haagy was ill with a serious lung condition and spasms of coughing that sapped his strength and confined him to his home in Pasadena. How did Haagy feel about progress on smog?

"We will always have some bad days, but there are so many more good days now."

Will the battle for clean air be won?

"If the goal is zero smog, we'll never reach it. There will always be some contamination. But our air on the average is fifty percent cleaner than it was in 1970. We have made progress." Haagy said, "The battle will go on."[152]

Less than two weeks later, on Friday, March 18, 1977, Haagy died. His death was big news for the *Los Angeles Times*, which reported it on page 1: "Caltech's Haagen-Smit, Smog Study Pioneer, Dies—Dr. Arie Haagen-Smit, the first man to analyze and trace the main sources of air pollution, died today of lung cancer. He was 76....Dr. Haagen-Smit broke down air pollution and became the so-called 'Father of Smog.'"[153]

The late Larry Pryor, professor at the University of Southern California's journalism school and former *Los Angeles Times* reporter and editor, recalled when he was a reporter covering the Air Resources Board about 1968. Haagy was its first president. "Having someone with the scientific authority be able to say, 'This is smog and this is how it is created and this is what it is going to take to fix [it]' right there in one man and in the influence he had over the other board members was incredible."

Pryor emphasized the importance of Haagy's turning point. "If we hadn't had Arie Haagen-Smit, if we hadn't had the Air Resources Board, if we hadn't had the controls that were placed on the auto industry back in the early 1970s, imagine what the air would be like today. We literally wouldn't be able to breathe."[154]

Beginning in 2001, as an incentive to encourage research and inspiration, the California Air Resources Board began to award each year the Haagen-Smit Prize for extraordinary accomplishments in the battle to improve air quality and climate change. That award, some say, has become recognized as the "Nobel Prize" in air quality achievement.

Geography and weather that create the temperature inversion that stagnates air over Los Angeles means the community likely must fight forever to win the war against air pollution. But no longer is Los Angeles the clear winner of the title Smog Capital of the World.

Today, the city of Los Angeles thrives because it has "air you can breathe." There's still smog. But it is not the thick, oppressive, silver-blue photochemical smog that threatened to drive everyone out of Los Angeles. Today's smog in Los Angeles is milder and brown. "The chemistry has changed," said Haagy's research associate Margaret Fox Brunelle.

"I remember when I was growing up…days when the smog was so thick that, standing at the front door of my home, I could not see the house across the street." That does not happen in Los Angeles anymore.

"I remember when I was growing up…days when smog was so heavy we were not allowed to play outside at school or at home." That does not happen in Los Angeles anymore.

For sixty years, Los Angeles earned fame as the Smog Capital of the World. But its last Stage 3 smog—the worst—struck in 1974, three years before Haagy died.[155]

Fourteen years later, the last Stage 2 smog struck in 1988.[156]

After another fifteen years, the last Stage 1 smog struck in 2003.[157]

Los Angeles has changed "air you can *see*" into "air you can *breathe*."

Appendix A

TUCKER REPORT, 1947

The Los Angeles Smog Report
by Prof. Raymond R. Tucker
Head of the Mechanical Engineering Department of
Washington University, St. Louis, and former Smoke
Commissioner of St. Louis

NOTE

The following "Smog Report" is reprinted for free distribution by the *Los Angeles Times* as a public service. The *Times* in December, 1946, brought Prof. Raymond R. Tucker, St. Louis smoke and fumes authority, to this city for a comprehensive survey of the source of so-called "smog"—smoke, and eye-irritating fumes—and then on Jan. 19, 1947, printed his recommendations.

It is hoped that the wide distribution of this booklet will lead to a better understanding of the problem involved, and will result in full public cooperation to put into effect the necessary remedies.
/s/ Norman Chandler
PUBLISHER
LOS ANGELES TIMES

The undersigned consultant was employed by the *Los Angeles Times* to investigate possible sources of atmospheric pollution in the city of Los Angeles and Los Angeles County which may contribute to this problem.

The results of this investigation were to be a guide for the formation of a program which would result in the eventual elimination or control of these sources to a point where they would cease to be a nuisance.

The type of smog visitations in the Los Angeles area is adequately described by the following quotation from a report addressed to D. Uhl (Los Angeles City Health Officer) dated July 30, 1943:

"On Sept. 21, 1942, the abnormal atmospheric conditions re-occurred. Although the downtown area was affected severely, the outlying districts were affected to a greater extent than at any previous time. At this time the quantity of smoke was greater since the industrial plants were operating longer hours. The industrial hygienists determined that the following irritants were in the air (especially true in the vicinities of the plants producing them): ammonia, formaldehyde, acrolein, acetic acid, sulphuric acid, sulphur dioxide, hydrogen sulphite, mercaptans, hydrochloric acid, hydrofluoric acid, chlorine, nitric acid, phosgene, and certain organic dusts known to be irritants. Some of the industrial plants emitting these irritants through furnace, stacks, vents, or by natural ventilation were as follows: foundries, oil refineries, chemical manufacturing plants, fish canneries, cinerators, hydroplating plants, fertilizer plants, packing plants, soap factories and waste disposal plants."

Apparently this problem became acute during the past five years, although there are some who contend that it was present many years ago. Reports in the possession of the Health Department, however, point to the past five years. Records of the Weather Bureau over the past 10 years show that the total annual mean visibility was a maximum in 1939. From 1939 until 1943 there was a rapid decline, with a minimum being reached in 1942. This same trend was apparent in the semiannual curves with a minimum reached in the same year. It would appear, therefore, that something had occurred during this period to cause this reduction in visibility.

In the industrial district of Los Angeles there are approximately 13,000 industries. Of these 13,000 about 1000 are in the unincorporated areas and 8000 to 9000 in the city of Los Angeles proper, the rest being in the incorporated areas.

A representative of the Chamber of Commerce estimated an increase in the number of industries and expansion of industries in the city of Los Angeles during this five-year period to be almost 85 per cent. A review of the statistical record of Los Angeles County development published by the Los Angeles Chamber of Commerce shows that the period of greatest industrial development in a monetary way was in the years 1942 and 1943, the total capital investment being approximately $261,000,000. This exceeds by $88,000,000 the amount invested from 1931 to 1940. It further represents 51 per cent of the total of $513,000,000 invested during the five-year period 1941 to 1946.

A large percentage of these industries, according to this same authority, are concentrated within a 10-mile radius to the north, the east and the south of a focal point located at Olympic Blvd. and Main St.

A report of the Health Department dated June 3, 1946, titled, "Progress Report—Smoke and Fumes Abatement," quotes Maj. Robert Dickman, their meteorologist, as follows:

"The prevailing wind is from the northeast during the months from November through March, changing to southwest and west during the remaining months. The average wind velocity is six miles per hour; however, a tabulation of wind directions for each hour between 8 o'clock a.m. and 5 o'clock p.m. for an entire year shows a prevailing wind from the southeast for the hours between 8 o'clock a.m. and 12 o'clock noon."

As the winds in the morning are from the above directions, the fumes, smoke, odors and dust are carried into the heart of the city from the industrial district. The major portion of these industries are of such character that they are capable of producing obnoxious gases, fumes, odors and dust. A review of the various types shows 41 different categories. Included among them are chemical industries, refineries, food product plants, soap plants, paint plants, building materials, nonferrous reduction refining and smelting plants, as well as numerous others of similar type. From the stacks of these are discharged sulphur dioxide, smoke, dust, aldehydes and other noxious gases.

These plants until recently had little or no supervision with reference to their discharged gases. As a result, they offer their share daily to the pollution of the atmosphere. These should be supervised and controlled.

They should be required to file with the Office of Air Pollution an analysis of their atmospheric discharge showing the per cent of all gases permitted to escape into the air, as well as the quantity of product which is the source of these gases.

This rapid increase in industry without any control over its operation and design contributed its share to the atmospheric conditions now existing in Los Angeles. They added to the many other existing sources which may cause the irritation of the eyes.

During the period 1940 to 1945 the population of the city of Los Angeles increased from 1,504,000 to 1,877,000 and the population of Los Angeles County from 2,785,000 to 3,703,000—an increase of 373,000 in the city and 918,000 in the county. It is common practice for the citizens of Los Angeles to burn all combustible refuse from the homes (on order from the Fire Department) between the hours of 8 a.m. and 10 a.m. This refuse is burned in many cases in the open and in a manner conducive to the production of smoke, fumes and irritants. The extent to which this occurs depends upon the nature of the material as well as the methods employed.

It is true that individually they mean nothing; but collectively they are a recognizable source of air pollution. As stated, individually they mean very little as a source of pollution; however, with a population of 1,877,000 it is estimated there are approximately four pounds per capita per day burned, or 3750 tons, in the city of Los Angeles every 24 hours. This would amount to 7400 tons for the county. Thus it is not one home that is involved, but the total number of homes.

With this viewpoint, it can be readily seen that the pollution resulting from these homes also represents an important position in the ultimate problem. The practice of burning rubbish should be discontinued unless done in properly designed incinerators.

Spread throughout the community are commercial and municipal dumps. When observed, these dumps were on fire and the smoke, odors and fumes were quite noticeable. The information available indicated that these dumps burned 24 hours a day with varying intensity. This constant burning of combustible refuse is certainly a contributing factor to atmospheric pollution. These dumps add their noxious gases to those from other sources. This problem was recognized and there is available a report dated July, 1945, and signed by the Refuse Disposal Committee outlining methods for the elimination of this source of nuisance.

Further investigation revealed the presence of many incinerators of improper design used for community disposal of rubbish or for the incineration of offal from manufacturing processes such as wood, etc. Aldehydes can be produced by the burning of wood. These incinerators, as observed, almost without exception are improperly designed and could not burn anything efficiently. These incinerators should not be used. Furthermore, only those incinerators whose design has been approved by the Office of Air Pollution should be allowed. With the increase in industry and population, the quantity of waste materials burned in this manner has greatly increased.

There are some who blame the present conditions on the automobile. A traffic survey submitted by the Automobile Club of Southern California indicates the number of automobiles crossing Figueroa St. between Sunset

Blvd. and Pico St. between 6 a.m. and 10 p.m. are as follows for the following months and years:
1936—159,015
October 1941—158,639
July 1945—136,700
January 1946—177,331

An inspection of these figures shows that between 1941 and 1944 there was a decrease in the automobiles entering this area of Los Angeles; however, it was in this period that the lachrymatic effect became most noticeable to the general public; in fact, the number of automobiles in January 1946, over October 1941, has only increased 11.7 per cent. It is not the intention of this report to absolve anyone from his responsibility. It would appear, however, that although it is quite probable that the automobile does contribute to the nuisance, it is not in such proportion that it is the sole cause. (Caution should be exercised in placing the entire blame on any one industry, plant or group of individuals. Each contributes its share.)

According to the figures available as of Dec. 9, 1946, the Los Angeles Transit Lines operated 429 buses. The P.E. operates for its Los Angeles passenger service 355 buses. There are those who claim that these 764 [*sic*] buses are the source of the trouble; however, in the year 1943 when public attention was focused on the lachrymatic gases, these same companies operated 635 buses. As in the case of the private automobile, the buses, no doubt, contribute to the nuisance but not in the manner some would lead us to believe. The following experience illustrates this point.

One plant executive admitted the discharge of 50 tons of SO_2 into the atmosphere every 24 hours. At the time of this admission, he stated that this quantity was nothing compared to the discharge from the automobile. It is assumed he meant the discharge of sulphur dioxide. Simple calculations reveal the fact that the 177,000 automobiles [coming in and out of Los Angeles] in 16 hours only discharge as a maximum two tons of SO_2. Thus one plant discharges approximately 25 times as much as all the automobiles coming in and out of the city of Los Angeles across Figueroa St.

Another offender and contributor to the general pollution of the atmosphere is the diesel truck. It is estimated that 7100 of these trucks enter and leave Los Angeles every day. Many are in need of repair and others are carelessly handled. There are some who disregard the manufacturer's specifications with reference to fuel feed and others who intentionally increase the fuel charge to the engine. This all results in the discharge of smoke and fumes to the atmosphere. Furthermore, these discharges are not limited to one place but are discharged to large areas along the routes of the trucks.

It is the practice during the winter months to use oil for heating in the larger buildings and commercial establishments. These oil burners if improperly installed, if poorly regulated, or if in need of repair will contribute their share to the aldehydes, noxious gases and fumes.

The railroads, too, may be a source of aldehydes and soot. The practice of sanding tubes on oil-fired locomotives usually results in a violent discharge of soot.

Los Angeles has another factor which accentuates this concentration of foreign material in the atmosphere. As we are all aware, Los Angeles County is situated between the Pacific Ocean, which is relatively cool the year round, and desert, which is hot in the summer and cool in the winter. The result is a monsoon-type of climate with prevailing onshore flow in the winter months. The free interchange of oceanic and continental air is hindered considerably by the coastal mountain range.

During the summer months, air moving eastward across the Central Pacific Ocean frequently turns sharply southward west of San Francisco, acquiring accelerated velocities at the turning point. Rapid southward movement of the Pacific air causes upwelling of cold water off the Central California coast, which cold water, in turn, cools the overlying layer of air; but there is a sufficient wind to keep this air well stirred and a normal lapse rate of temperature persists in the low levels.

In other words, throughout the depth of the cool but mechanically stirred marine layer, temperature decreases with height at the rate of about 4 deg. F./1000 feet. As this shallow layer of air moves southward, it continues to be heated in the lower levels and if enough moisture is added by evaporation, condensation will take place at the top of the marine layer and a stratus cloud or fog will form. As this layer of air moves inland, with the sea breeze on the Los Angeles coast, daytime heat may be sufficient to establish convection up through the stratus cloud and cause it to dissipate. After sunset there is a tendency for the lower levels of the atmosphere to be cool and stratus or high fog will re-form near the original condensation level.

The above is a rather simple explanation of the typical summer high fog regime in the Los Angeles area. Such a situation occurs when the interior desert region has high temperatures and low pressure. The lower the pressure in the interior as compared to the coast, the more pronounced will the onshore flow be and to a limited extent, the deeper will be the marine layer.

During and preceding a hot spell when pressures along the coast become relatively low (such as on Friday, Sept. 13, 1946), pressures in the interior become relatively higher and the warmer continental air moves westward toward the coast. In such cases, the marine layer of air becomes very shallow and the top of this layer, marked by a temperature inversion, gets very close to the ground. Such a low temperature inversion is usually associated with very stagnant air in the low levels.

Such conditions tend to persist two to five days accompanied by cumulative pollution of the very lowest layers of atmosphere over Los Angeles County. Coastward migration of desert air also produces very pronounced temperature inversions which are difficult to dissipate by daytime heating. When the land becomes very hot, temperature difference alone may be responsible for development of a weak sea breeze which tends to undercut the hot desert air in a very shallow layer.

For example, Friday, Sept. 13, 1946, Burbank, at an elevation of about 600 feet over the city of Los Angeles, had a maximum temperature of seven degrees higher than the downtown area. A very shallow sea breeze struggles landward on such occasions and even though it has dissipated at its leading edge quite rapidly, it is strong enough to keep the lowest layer of air cool and causes the temperature inversion to persist. This sea breeze migrates inland and up the river channels reaching Glendale about 10 and Burbank about noon, bringing with it much lower visibility and frequently its arrival in Burbank will be accompanied by considerable irritation of the eyes. The sea-breeze is a normal almost daily phenomenon, but is not always accompanied by irritation of the eyes.

Samples of recent records of the occurrence of strata, high fog and low temperature inversions indicate that the last two years have shown no greater frequency of any of these phenomena than could reasonably be expected from available compilations of frequency of these elements. The seven years of statistics used as a normal yardstick were compiled during the period 1932 to 1938, inclusive, a period which antedates the smoke nuisance.

This atmospheric condition is, therefore, not of recent date, so whatever has occurred cannot be blamed upon the atmosphere, although it may be accentuated by atmospheric conditions.

Thus we have in the Los Angeles area industry, individual refuse fires, commercial and municipal incinerators, dumps, private incinerators, buses, diesel trucks, railroads and heating plants contributing to air pollution. It might be truly said that these conditions existed at all times and it is only recently that conditions have become unbearable. However, there is a limit to the pollution the atmosphere can absorb without becoming noticeable. With the increase of industrial activity, with the increase in population, the discharge of foreign material into the atmosphere has also increased. The concentration of these gases, fumes and irritants has reached the threshold of human sensitivity. It would not be reasonable to expect any other results.

As originally pointed out, these conditions have become aggravated during the past five years. From data available, no new or unusual event has occurred to which could be traced the source of gases which are causing the lachrymatic effect.

At one time it was thought that the butadiene plant was a sole source of the irritants. When this plant corrected its conditions, the eye irritants still persisted. Chemical analysis of the area is not very revealing. Although irritants are identified, they are not of sufficient quantity, according to published data, to cause the present effect.

Many theories have been advanced as to the cause, but as yet none have been substantiated. The research work to ascertain and identify the gas or gases producing this effect is under the control of the county. Mr. I.A. Deutch has ably planned this work. The eventual solution will be had. In the meantime, unless drastic steps are taken to curtail and eliminate all visible stack discharges, conditions will become worse.

As the ordinances of both the city and the county are reviewed, one is impressed by the lack of authority possessed by enforcement officers. They should occupy a separate and distinct place in the city and county organization. They should be recognized not only as the nominal head of the division but as the actual head. Their authority should flow in an unbroken line from the legislative body. They should be the sole arbiter in matters pertaining to air pollution, subject, of course, to review by the courts or an appeal board.

They should not be hampered by restrictive legislation such as Section 731-A of the Code of Civil Procedure, which for all practical purposes legalizes the creation of a nuisance. They should be empowered by the legislative body with sufficient authority to cause the discontinuance of use of equipment, if necessary. It is foolish to prohibit the creation of a nuisance if the enforcement officer is not given sufficient power to enforce the terms of the act.

Fundamentally, the hands of Maj. Harry E. Kunkel and Mr. I.A. Deutch are tied. They do not have the necessary authority to proceed in an aggressive and logical manner. Maj. Kunkel is operating under an ordinance which has been recently passed. This ordinance merely declares that smoke, dust and fumes, etc., are a nuisance if they exceed a density of 40 per cent. If less than 40 per cent they may pollute the atmosphere for 24 hours every day. Mr. Deutch is operating under a similar ordinance. [It is more liberal and permits smoke and fumes of less than 60 percent density to be continually discharged.]

The discharge of gases, fumes, soot and smoke should only be permitted for the specified times of three minutes when cleaning the equipment or in starting the equipment.

The county ordinance only applies to the unincorporated areas, while many of the larger industries are located in incorporated areas. Fourteen incorporated areas out of more than 40 have agreed to permit county supervision of their smoke problem. All the incorporated and unincorporated communities must be subjected to the same supervision, because it has been wisely said that smoke, fumes and dust are not conscious of jurisdictional or boundary disputes.

The present condition of divided authority, viz. the city, county and incorporated areas, can practically nullify the efforts of any enforcement group no matter how competent or energetic. The city ordinance does not permit the Director of Air Pollution to supervise the design or installation of any combustion equipment or stack from which fumes, smoke, dust, etc., may be discharged. Thus the director may correct one installation and have four others installed which will be a prolific source of contamination. This all can be prevented if proper authority were had.

The county, on the other hand, does require installation permits and operating certificates. This authority is not placed in the hands of the Director of Air Pollution but in the office of the chief engineer of the Building Department. No matter how competent this gentleman may be as a structural engineer, he is not competent to pass [judgment] upon combustion problems or equipment. He, no doubt, refers these problems to the Air Pollution Department. If he does, why not do it directly by law?

It is further noted that the regulations, as published, are not comprehensive. There are specific requirements for some conditions and others are decided as individual cases. It would be far better to be less specific and only include in the ordinance the authority to promulgate rules and regulations by the director. The director would then be in a position to amend these rules as experience dictated and include new specifications as the need arose, without requiring the passage of a new ordinance.

Any air pollution program must attack the nuisance at the source. The atmosphere cannot be controlled, but the discharge of contaminants into the atmospheric air can be controlled. To accomplish this, all fuel and equipment being used and all the equipment being used and all the equipment installed, repaired, or remodeled must be controlled by the Division of Air Pollution. To do this work properly requires trained and competent personnel. This personnel should have no other duties other than air pollution control. They cannot be health officers one day and air pollution officers the next. Control of air pollution is a full time job. It is continuous and perpetual.

Fortunately the city of Los Angeles and Los Angeles County have two very able and competent men in immediate charge of the present program. They alone, with inadequate tools, cannot solve the problem. Even with adequate tools, they alone could not solve the problem. They need public support as well as adequate legislation. Mere passage of an ordinance does not eliminate air pollution; it only authorizes some individual or group of individuals to clarify the air. If new legislation is had, the citizens must be made to realize the necessity of their sustained interest in the program. The elimination of smoke in any community is a dual responsibility which should be shared by public officials and citizens alike. Public officials are not miracle men. They cannot wave a wand and clear the atmosphere. It takes time, work and the expenditure of money to obtain the results desired. The citizens may have as clean an atmosphere as they desire. It is up to them to decide. This consultant hopes to point the way. The citizens themselves must make the ultimate decision This they should remember… you cannot continue to pollute the atmosphere to ever-increasing amounts and not pay the penalty in dust, filth, fumes, physical discomfort and impaired health.

As the sources of lachrymatic gases have not been determined, there are two methods of approach to the solution of this problem in the Los Angeles area, the first being to intensify and accentuate the research now being carried on to identify the gas, gases or solid material that is causing eye irritation.

This work may require only a few months or it may require many years. If the solution of the problem is had, it then becomes mandatory to go back to the source of all visible evidence of pollution in order to determine the quantity of this particular element being discharged by each stack or vent. The question then arises as to the limitation or the minimum quantity to be discharged. So, in the final analysis, the possible sources of air pollution must be sought and controlled.

This intensified research program could be carried out while operating under present inadequate ordinances. Its results, however, will be unsatisfactory because the possible contaminants would be increasing from all those sources outside the jurisdiction of the enforcement officer.

The other method is to rigidly control all known sources of pollution under proper and comprehensive legislation while carrying on the research program. With this method, conditions will be prevented from becoming worse and the enforcement officers can backtrack and reduce all possible sources of pollution now in existence. This cannot be accomplished by voluntary cooperation nor can it be accomplished if anyone is granted immunity from any provision of the regulations. The smallest incinerator to the largest industry, as well as the railroads and trucks, must conform to the requirements established.

The adoption of the second method appears to be sound. The elimination of all visible discharge into the atmosphere of dust, fumes and gases must be done if atmospheric conditions are to be improved.

In this necessary reduction of contaminants, it is quite possible that the illusive element causing the irritation of the eyes may be reduced below the threshold of human sensitivity. Furthermore, during those periods of prolonged temperature inversions, the concentration of the entrapped gases will be reduced, thus again affording a possible relief from the conditions now existing.

It is the conclusion of this consultant, from a review of the existing conditions in the Los Angeles area, that the following recommendations should be followed:

1—That the second method of attack on this problem be adopted.
2—That the necessary State legislation be enacted to create an air pollution control district, preferably county-wide. This legislation should be of such character as to enable the enforcement of regulations in all areas of the county, unincorporated and incorporated.
3—That in the creation of this air pollution district, broad, general powers be given to the district, if in accordance with California law, so as to permit them to enact rules and regulations pertinent to their particular problems. No standards as to what is dense smoke or a nuisance should be set out in the law because what may be objectionable in one community may not be in another. The line of authority should flow from the Legislature to the county and from the county to the enforcement officer.
4—That Section 731-A of the Code of Civil Procedure be repealed or amended to remove any doubt as to the authority of the enforcement officer.

5—That the present city ordinance be amended so that all fuel-burning equipment being used and all equipment installed, repaired or remodeled come under the control of the Director of Air Pollution.
6—That the present city ordinance be amended so that the Director of Air Pollution has authority to seal the equipment of persistent violators.
7—That adequate personnel be assigned to the present city Office of Air Pollution for work on air pollution alone.
8—That the present city and county ordinances be amended so that the emission of No. 2 smoke on the Ringelmann chart be permitted only when a new fire is being started or during cleaning periods.
9—That the present county ordinance be amended to change the supervision of all fuel-burning equipment being used, installed, remodeled or repaired to the office of the Director of Air Pollution.
10—That the present county ordinance be amended so that the man who is actively in charge of air pollution is the actual director and not the assistant director, thus eliminating the possibility of divided authority.
11—That the present county ordinance be amended so that all specific requirements for installations are deleted. There should be inserted instead the authority for the director to promulgate rules and regulations for the guidance of the installers.
12—That the present ordinances be amended to require all known sources of air pollution from the stacks of industry, commercial establishments, etc., to file, upon the request of the Director of Air Pollution, a certified analysis of the products being discharged from their stacks or vents. This analysis should also contain a statement as to the quantity discharged.
13—That the present ordinances be amended requiring dust collectors or precipitators to be installed in connection with those devices which are discharging noxious gases and irritants into the air, unless other ways and means are used to reduce this discharge to a minimum.
14—That the present ordinances be amended to prohibit the burning of combustible refuse in municipal and commercial dumps. This may necessitate the building of private or public incinerators.
15—That the present ordinances be amended prohibiting the present practice of burning rubbish in the back yards of homes unless in an incinerator approved by the Director of Air Pollution. This may necessitate the public collection of all combustible refuse, to be subsequently burned in public incinerators.
16—That the present county ordinance be amended so that the Director of Air Pollution shall have the authority to seal the equipment of persistent violators.

17—That the present ordinances be amended prohibiting the use of present type of private and public incinerators used for the burning of offal from industry unless approved by the Director of Air Pollution.

18—That the present ordinances be amended to permit the stopping of diesel trucks at the time of emission of dense smoke and making it a misdemeanor to operate the trucks under any other conditions than those recommended by the manufacturer.

19—That the present ordinances be amended requiring the diesel buses and trucks to have periodical overhauls as recommended by the manufacturer and that each truck be required to carry a certificate showing the time and extent of that overhaul.

20—That the present ordinance be amended requiring the railroads to discontinue the use of oil-burning locomotives in switching and transfer service and to substitute in lieu thereof diesel locomotives.

21—That increased emphasis should be placed on the elimination or reduction of all known sources of atmospheric pollution.

22—That the present research program on the source of eye irritants be continued.

[Signed]

RAYMOND R. TUCKER

Consultant

Appendix B

LOUIS C. McCABE LETTER TO THE LOS ANGELES COUNTY BOARD OF SUPERVISORS, SEPTEMBER 13, 1948

Honorable Board of Supervisors
Gentlemen:

From time to time I have reported the activities of the [Air Pollution Control] District to you. You are aware of the headway we have made and are continuing to make in the control of air pollution from burning dumps, the ferrous and non-ferrous foundries, steel plants, industrial incinerators, trucks, locomotives, and improper combustion in general. It is appropriate that I bring you up-to-date on another phase of our work.

On August 14, we first collected the droplets (aerosols) which constitute the liquid phase of the smog. This was done with the cascade impactor which deposits the particles of different size on glass slides which are placed in front of the four jets of the apparatus. Since then we have continued this work and on three days of the past week we collected twenty-four hours continuously. Chemical tests show the material to contain sulfates and to be acid in character.

The greatest quantities of sulfur compounds are released to the atmosphere in the Los Angeles area from the refineries, chemical plants and the burning of fuel oil by other industries. The sulfur balance supplied me by the petroleum industry some time ago indicates that 822 tons of sulfur dioxide enter the Los Angeles atmosphere each day. Twenty-two tons of this comes from gasoline burned in automobiles and twelve tons are released by the burning of diesel oil.

The fog-forming qualities of SO_3 (sulfur trioxide) and sulfuric acid mist are well known. That sulfuric acid is formed by the oxidation of sulfur dioxide in sunlight has been demonstrated by German and American chemists. We have repeated these experiments here and find that if a flask filled with sulfur dioxide is left for as long as 20 minutes in the sunlight a fog of sulfuric acid mist is formed. The extremely low visibility that begins two or three hours after sunrise when the humidity is too low for natural fog is very likely due to this formation of sulfuric acid mist.

We have collected smog at stations set up at 223rd St., just east of Santa Fe, at 5201 Santa Fe Ave., 808 N. Spring St., and in the vicinity of the Rose Bowl. It is our observation that heavy concentration occurs earliest in the morning at the first station and moves on through the others in the order named as the southwest-to-west wind pattern is established.

The past two weeks has been a period of relative calm wind accompanied by temperature inversions with their base near the ground. This condition concentrates the smog in a thin layer along the ground. August 31 to September 4 was marked by extremely heavy pollution. About sixty per cent of the refining capacity in the area began closing down Sept. 3 and was in standby condition on Sept. 5. Although meteorological conditions during the week since the beginning of the strike are not the same as the week before, there have been days of low inversions. However, the visibility has been much better and the pollution has been less intense. The clarity of the atmosphere on the leeward side of the idle refineries is in striking contrast to that of the operating plants.

Our observations are that sulfur compounds contribute in great measure to the smog. The removal of sulfur gases from industrial stacks is by no means an easy problem to solve. A few of the refineries in the Los Angeles area now recover hydrogen sulfide economically and thus reduce the sulfur dioxide going to the atmosphere. Another chemical plant is scheduled for completion by January 1 which will process fifty tons of hydrogen sulfide daily, thereby reducing the amount of sulfur compounds in the atmosphere. However, I do not believe this goes far enough. Because of climatic conditions in the Los Angeles basin, we must apply strict standards comparable to those established by the London County (England) Council. We are continuing our work on sulfur and hydrocarbon compounds. At the same time we are developing standards which will be submitted to you for approval.

You know of our unfruitful attempt to secure a surplus ammonia plant for Southern California from the War Department. Cheap ammonia would make it possible to use the ammonium sulfate or the ammonium sulfite-bisulfite process to remove sulfur dioxide from stack gases. Ammonium sulfate for fertilizer and liquid sulfur dioxide produced by these processes should find a ready market. You can appreciate that if the War Department were to allocate a surplus ammonia plant to this area there is a relatively easy solution to a major part of our smog problem.

Respectfully submitted,
/Signed/
Louis C. McCabe
Director

Appendix C

LOUIS C. McCABE LETTER TO THE LOS ANGELES COUNTY BOARD OF SUPERVISORS, FEBRUARY 25, 1949

Honorable Board of Supervisors
County of Los Angeles
501 Hall of Records
Los Angeles 12, California
Gentlemen:

I wish to report a most significant new development in our investigation of the complex problems of air pollution. As a direct result of our recent research contract with the California Institute of Technology we have discovered that an important factor in eye irritation may be the presence of substances known as organic peroxides. Dr. A.J. Haagen-Smit, Professor of Bio-Organic Chemistry, California Institute of Technology, has found surprisingly large quantities of organic peroxides in samples collected by our Dr. Paul P. Mader. It is well known that organic peroxides are extremely irritating, but such compounds have not heretofore been reported as significant air pollutants. Our present preliminary work indicates, however, that these substances may well be the most important cause of eye irritation in this area.

Although these results are based upon analyses of only a few samples, and the investigation is continuing, it was decided at a conference held last Tuesday, attended by Dr. Haagen-Smit, Dr. Beckman and Dr. Blacet, who are consultants of the District, to report the preliminary findings immediately rather than wait for a final report because of the widespread public interest in the air pollution problem. We believe also it is in the public interest to bring these findings promptly to the attention of other technical groups interested in various phases of air pollution.

Sincerely yours,

/Signed/

Louis C. McCabe

Director

NOTES

Chapter 1

1. Gumprecht, *Los Angeles River*.
2. Bill Garnett, email to author, October 12, 2023; Meredith May, "William A. Garnett—Aerial Photographer, UC Professor," September 10, 2006, www.sfgate.com/bayare/article/william-a-garnett-aerial-photographer-uc-2552495.php.
3. Wagner, *Juan Rodriguez Cabrillo*.
4. Philip Hager, "Gravestone Believed to Be Cabrillo's Displayed at UC," *Los Angeles Times*, December 21, 1972.
5. Bill McAllister, "Discovering an Explorer," *Washington Post*, September 11, 1992.
6. Bottles, *Los Angeles and the Automobile*.
7. Zeder, "Million Dollar Search," 97.
8. Graves, *California Memories*.
9. Went, "General Problem of Air Pollution and Plants," 148–49.
10. "Exhaust Fumes Hang Over City," *Los Angeles Times*, July 23, 1943.
11. Senn, letter to Uhl.
12. Cadle and Wohlers, "Smog Lore," 30–35.

Chapter 2

13. California State Military Museum, "The Shelling of Ellwood," February 8, 2016, www.militarymuseum.org/Ellwood.html.
14. Tom Modugno, "The Sub Commander and the Cactus Myth, Debunked," February 28, 2021, goletahistory.com/the-sub-commander-and-the-cactus-myth-debunked.
15. California State Military Museum, "The Battle of Los Angeles," June 23, 2017, www.militarymuseum.org/BattleofLA.html.
16. "Los Angeles, First Center of Defense," *Los Angeles Times*, December 1, 1941.
17. "Don't Cry! Smarting Gas Not Chemical War," *Los Angeles Daily News*, July 27, 1943.
18. "Butadiene Plant Confirmed as Source of Gas Fumes Here," *Los Angeles Times*, July 29, 1943.
19. Los Angeles City Health Department, *Progress Report on Smoke Elimination*; "End Ordered for Gas Fume Annoyances," *Los Angeles Times*, September 19, 1943.
20. Senn, letter to Uhl.
21. Brienes, "Smog Comes to Los Angeles," 515–32.
22. "Fumes Storm Spurs Quest for Remedy," *Los Angeles Times*, October 6, 1943.
23. Ibid.
24. Ray Zeman, "City 'Smog' Laid to Dozen Causes," *Los Angeles Times*, September 18, 1944.
25. "Officials Clash on Fumes Plant," *Los Angeles Times*, October 17, 1943.
26. Bowron, letter to the Council of the City of Los Angeles.
27. "City May Drop Nuisance Suit on Gas Fumes," *Los Angeles Times*, October 27, 1943.
28. "L.A. Council Rescinds Suit on Butadiene Fumes," *Los Angeles Evening Herald-Express*, October 27, 1943.
29. Zeman, "City 'Smog' Laid to Dozen Causes."
30. Ed Ainsworth, "Synthetic Rubber Industry Shows It's Possible to Eliminate Smog," *Los Angeles Times*, November 4, 1946.

Chapter 3

31. Zeman, "City 'Smog' Laid to Dozen Causes."

32. Brimblecombe, *Big Smoke*.
33. Los Angeles, CA, Ordinance 89,893 Amending Ordinance 89,100, Salary Standardization Ordinance of the City of Los Angeles, November 19, 1945.
34. Kunkel, "Air Pollution Control."
35. Royce, *Life Remembered*.
36. "Smog Blanket Densest Here Since End of War," *Los Angeles Times*, September 14, 1946.
37. Nicholson, "Los Angeles Battles the Murk," 17–19, 90–93; Mitchell, "Clearing the Air."
38. Boyarsky, *Inventing L.A*.
39. Ed Ainsworth, "Fight to Banish Smog, Bring Sun Back to City Pressed," *Los Angeles Times*, October 13, 1946.
40. Los Angeles County Smoke and Fumes Commission, Report to the Los Angeles County Board of Supervisors.
41. "We Can Curb the Smog Nuisance," *Los Angeles Times*, July 20, 1946.
42. Raymond V. Darby, "County Is 'Ready, Willing and Able,'" *Los Angeles Times*, December 9, 1946.
43. Ed Ainsworth, "St. Louis Has Key to Smog," *Los Angeles Times*, November 9, 1946.

Chapter 4

44. Ed Ainsworth, "'Times' Bringing Smog Expert Here," *Los Angeles Times*, December 1, 1946.
45. I.A. Deutch, "Posing the Smog Problems," *Los Angeles Times*, December 9, 1946.
46. "Jeffers to Head Smog Committee," *Los Angeles Times*, December 20, 1946.
47. Ed Ainsworth, "Smog Fighter Ends Survey," *Los Angeles Times*, December 23, 1946.
48. "This Smog Fight Is No Picnic," *Los Angeles Times*, December 24, 1946.
49. Ibid.
50. William W. Larsen, "Real Optimist Heard From: He Finds Good in the Smog," *Los Angeles Times*, December 7, 1949.
51. "This Smog Fight Is No Picnic."
52. Ed Ainsworth, "'Times' Expert Offers Smog Plan." *Los Angeles Times*, January 19, 1947; Raymond R. Tucker, "Los Angeles' Smog Report," *Los Angeles Times*, 1947, reprinted for free distribution.

53. Tucker, "Los Angeles' Smog Report."
54. "The Smog Cleanup Begins," *Los Angeles Times*, January 26, 1947.

Chapter 5

55. *Air Pollution*, Proceedings of the United States Technical Conference on Air Pollution.
56. Chester G. Hanson, "Antismog Measure Passes Assembly by 73–1 Vote," *Los Angeles Times*, May 8, 1947.
57. Daniel L. Royce, letter to the author, March 13, 1995.
58. Royce, *Life Remembered*.
59. Ibid.; Kennedy, "History, Legal and Administrative Aspects of Air Pollution Control."
60. Royce, *Life Remembered*; Kennedy, "History, Legal and Administrative Aspects of Air Pollution Control."
61. Royce, *Life Remembered*; Kennedy, "History, Legal and Administrative Aspects of Air Pollution Control"; Koster, "History of Air Pollution Control Efforts."
62. Ed Ainsworth, "Smog Controller McCabe Arrives to Assume Duties," *Los Angeles Times*, October 2, 1947.
63. Ed Ainsworth, "New Trash-Burning Hours Hinted as Possible Smog Evil Solution," *Los Angeles Times*, October 7, 1947.
64. McCabe, "Text of M'Cabe's Report on Smog."
65. Ed Ainsworth, "Center for Burning Junk Worst Smog Offender," *Los Angeles Times*, September 29, 1947.
66. Ed Ainsworth, "Plants Help Smog Battle," *Los Angeles Times*, September 30, 1947.
67. Ed Ainsworth, "Smog Abatement Problem Posed by Small Industries," *Los Angeles Times*, October 3, 1947.
68. Ed Ainsworth, "Iron Foundries Add to Smog Nuisance," *Los Angeles Times*, October 4, 1947.
69. Ed Ainsworth, "What to Do About Smog Poses Problem for County," *Los Angeles Times*, October 5, 1947.
70. Ed Ainsworth, "Plant Shows How to Cut Dust Peril in Smog Drive," *Los Angeles Times*, October 6, 1947.

Chapter 6

71. McCabe, "Annual Report 1947–48."
72. Beckman, "Remarks at the John and Alice Tyler Ecology Award Dinner."
73. Ibid.
74. McCabe, Letter to Los Angeles County Board of Supervisors, September 13, 1948.
75. McCabe, "Text of M'Cabe's Report on Smog."
76. Ed Ainsworth, "Refineries Held Smog Leaders," *Los Angeles Times*, September 15, 1948.
77. Ibid.
78. Ibid.
79. "'Not So,' Oil Industry Replies to Smog Detectives' Findings," *Los Angeles Times*, September 28, 1948.
80. Beckman, "Remarks at the John and Alice Tyler Ecology Award Dinner."
81. "Smog's Complex Make-up Stressed," *Los Angeles Times*, December 9, 1949.
82. Beckman, "Remarks at the John and Alice Tyler Ecology Award Dinner."
83. Brechler, "Bouquet to 'the Father of Smog,'" 7–11.
84. Maria "Zus" Haagen-Smit, interview by author, July 19, 1994, Pasadena, CA.

Chapter 7

85. Haagen-Smit, "Power of Microanalysis," 496–99.
86. Brechler, "Bouquet to 'the Father of Smog,'" 7–11.
87. Beckman, "Remarks at the John and Alice Tyler Ecology Award Dinner."
88. Brechler, "Bouquet to 'the Father of Smog,'" 7–11.
89. Beckman, "Remarks at the John and Alice Tyler Ecology Award Dinner."
90. Haagen-Smit, "Light Side of Smog," 330–35.
91. Maria "Zus" Haagen-Smit, interview by author.
92. Haagen-Smit, "Smell and Taste," 28–32.
93. Schiller, "Los Angeles Smog," 558–64.
94. Eckels, "Los Angeles Pioneers in the Fight Against Smog," 30–34.
95. Haagen-Smit, "Light Side of Smog," 330–35.
96. Fisher, "Haagen-Smit on Smog," 36–39.

97. Margaret F. Brunelle, interview by author, March 3, 1996, Studio City, CA; Maria "Zus" Haagen-Smit, interview by author; McCabe, "National Trends in Air Pollution," 50–54.
98. Beckman, interviewed by Mary Terrall.
99. Beckman, "Remarks at the John and Alice Tyler Ecology Award Dinner."
100. Brechler, "Bouquet to 'the Father of Smog,'" 7–11.
101. Haagen-Smit, "Chemistry and Physiology of Los Angeles Smog," 1342–46.
102. Brechler, "Bouquet to 'the Father of Smog,'" 7–11; Brunelle, interview by author.
103. Brunelle, interview by author.
104. Stephens, Hanst, Doerr and Scott, "Reactions of Nitrogen Dioxide and Organic Compounds in Air," 1498–1504; Johnstone, McCabe and Thomas, "Chemistry of Pollutants in the Atmosphere," 1483.
105. Haagen-Smit, "Air Pollution Problem in Los Angeles," 7–13; Haagen-Smit, "Smog Research Pays Off," 11–16.
106. Brechler, "Bouquet to 'the Father of Smog,'" 7–11.
107. Haagen-Smit, "Smog Research Pays Off," 11–16.
108. Haagen-Smit, "Light Side of Smog," 330–35.
109. Al Martinez, "Think of Him When Sky Is Blue, Air Sweet…," *Los Angeles Times*, March 6, 1977.

Chapter 8

110. Haagen-Smit, "Light Side of Smog," 330–35.
111. Haagen-Smit, "Smog Control," 9–14.
112. Haagen-Smit, "Air Pollution Problem in Los Angeles," 7–13.
113. Haagen-Smit, Brunelle and Haagen-Smit, "Ozone Cracking in the Los Angeles Area," 1134–42; England, Krimian and Heinrich, "Weather Aging of Elastomers on Military Vehicles," 1143–54.
114. Tom Dunigan, "Air We Breathe Harms Tires on Cars We Drive," *Los Angeles Times*, February 1, 1959.
115. Ford, *Thirty Explosive Years in Los Angeles County*.
116. Eckels, "Los Angeles Pioneers in the Fight Against Smog," 30–34.
117. Lodge, "A.J. Haagen-Smit" [obituary], 565–66.

Chapter 9

118. Hahn, *Air Pollution—1967.*
119. Ibid.
120. "Automotive Engineer Group Arrives for Smog Checkup," *Los Angeles Times*, January 26, 1954.
121. Heinen and Fagley, *Smog.*
122. Ibid.; Campbell, "Instrumentation and Measurement Problems."
123. Campbell, "Instrumentation and Measurement Problems."
124. Haagen-Smit, "Light Side of Smog," 330–35.
125. Zeder, "Reduction of Exhaust Emission."
126. Dewey, "Antitrust Case of the Century," 341–76.
127. Martinez, "Think of Him When Sky Is Blue."
128. Gladys Meade, interview by author, Torrance, CA, March 4, 1996.
129. Martinez, "Think of Him When Sky Is Blue."
130. Lillard, *Eden in Jeopardy*, 242–43; Haagen-Smit, "Sins of Waste," 16–19, 30–32.
131. Fisher, "Haagen-Smit on Smog," 36–39.
132. Robert Fairbanks, "Moretti's Two Nonpolluting Steam Cars Make Trouble-Plagued Debut," *Los Angeles Times*, May 22, 1974.
133. Martinez, "Think of Him When Sky Is Blue."
134. Fisher, "Haagen-Smit on Smog," 36–39.
135. Martinez, "Think of Him When Sky Is Blue."
136. Ibid.
137. Ibid.
138. Bonner, "Arie Jan Haagen-Smit," 28–29.

Chapter 10

139. "California Maps Attack on Smog," *New York Times*, June 14, 1959.
140. "New Type Exhaust May End Monoxide Danger," *Los Angeles Times*, December 17, 1950; "Smog Muffler Found in Need of Improvement," *Los Angeles Times*, May 10, 1955.
141. *Motor Vehicles Industry Efforts to Reduce Air Pollution from Exhaust.*
142. Sperling and Dill, "Unleaded Gasoline in the United States," 45–52.
143. Ibid.
144. *Price Differential Between Leaded and Unleaded Gasoline.*

145. Larry Pryor, "Fighting Air Pollution: It's Everybody Else's Business," *Los Angeles Times*, July 14, 1974.
146. W.B. Rood, "Experts Take Dim View of Clean Air Outlook," *Los Angeles Times*, December 8, 1975.
147. California Air Resources Board, "History."
148. Brimblecombe, *Big Smoke*.
149. Louis Fleming, "Air Pollution Menaces Richest Farm Areas," *Los Angeles Times*, January 8, 1961.
150. "Europe Cities Barring Cars to Fight Smog," *Los Angeles Times*, December 26, 1972.
151. Newhouse News Service, "Greece Takes Steps to Save Statues," *Los Angeles Times*, June 27, 1976.
152. Martinez, "Think of Him When Sky Is Blue."
153. "Caltech's Haagen-Smit, Smog Study Pioneer, Dies," *Los Angeles Times*, March 18, 1977.
154. Rosenberg, "How Los Angeles Began to Put Its Smoggy Days Behind."
155. South Coast Air Quality Management District, "Upland, Calif., Had Last Stage III Smog Alert in U.S."
156. South Coast Air Quality Management District, "Historic Ozone Air Quality Trends."
157. South Coast Air Quality Management District, "Southland's First Stage One Episode in Five Years."

SELECTED BIBLIOGRAPHY

Air Pollution. Proceedings of the United States Technical Conference on Air Pollution. New York: McGraw-Hill, 1952.

Beckman, Arnold O. Interviewed by Mary Terrall. October 16–December 4, 1978 (Pasadena: California Institute of Technology Archives). resolver.caltech.edu/CaltechOH:OH_Beckman_A.

———. "Remarks at the John and Alice Tyler Ecology Award Dinner, Beverly Hills, CA." Unpublished manuscript, May 3, 1979.

Bonner, J. "Arie Jan Haagen-Smit." *Engineering and Science* (May–June 1977): 28–29.

Bottles, Scott L. *Los Angeles and the Automobile*. Berkeley: University of California Press, 1987.

Bowron, Fletcher. Letter to the Council of the City of Los Angeles, October 13, 1943, Los Angeles City Archives Council Communications, Box A-832, vol. 3588, 15399.

Boyarsky, Bill. *Inventing L.A.: The Chandlers and Their Times*. Los Angeles: Angel City Press, 2009.

Brechler, C. Carlton. "A Bouquet to 'the Father of Smog." *General Motors Quarterly* (Winter 1974).

Brienes, Marvin. "Smog Comes to Los Angeles." *Southern California Quarterly* 58, no. 4 (1976): 515–32.

Brimblecombe, Peter. *The Big Smoke*. London: Methuen, 1987.

Cadle, Richard D., and Henry C. Wohlers. "Smog Lore." *Air Repair* 1, no. 4 (1952): 30–35.

California Air Resources Board. "History." ww2.arb.ca.gov/about/history.

Campbell, John M. "Instrumentation and Measurement Problems." In *Motor Vehicle Industry Efforts to Reduce Air Pollution from Exhaust, Papers Delivered at National West Coast Meeting, Society of Automotive Engineers, Seattle, WA*, no. 168, August 16, 1957.

Dewey, Scott H. "The Antitrust Case of the Century." *Southern California Quarterly* 81, no. 3 (n.d.): 341–76.

Doyle, Arthur Conan. "The Adventure of the Bruce-Partington Plans." *The Complete Sherlock Holmes*. New York: Doubleday, 1930.

Dreyfuss, Henry. *The Symbol Sourcebook: A Comprehensive Guide to International Graphic Symbols*. New York: McGraw-Hill, 1972.

Eckels, Richard P. "Los Angeles Pioneers in the Fight Against Smog." *The Reporter* 11, no. 12 (December 30, 1954): 30–34.

England, W.D., J.A. Krimian and R.H. Heinrich. "Weather Aging of Elastomers on Military Vehicles." *Rubber Chemistry and Technology* 32, no. 4 (1959): 1143–54.

Evelyn, John. *Fumifugium.* London: W. Godbid, 1661. Reprinted by the National Society for Clean Air. Dorchester, UK: Dorset Press, 1961.

Fisher, Dan. "Haagen-Smit on Smog." *Westways* (August 1972): 36–39.

Ford, John Anson. *Thirty Explosive Years in Los Angeles County*. San Marino, CA: Huntington Library, 1961.

Graves, Jackson A. *California Memories (1857–1930).* Los Angeles: Times-Mirror Press, 1930.

Gumprecht, Blake. *The Los Angeles River.* Baltimore, MD: Johns Hopkins University Press, 1999.

Haagen-Smit, Arie J. "The Air Pollution Problem in Los Angeles." *Engineering and Science* (December 1950): 7–13.

———. "Chemistry and Physiology of Los Angeles Smog." *Industrial and Engineering Chemistry* 44, no. 6 (1952): 1342–46.

———. "The Light Side of Smog." *Chemtech* 2, no. 6 (June 1972): 330–35.

———. "The Power of Microanalysis." *Journal of Chemical Education* 28 (1951): 496–99.

———. "The Sins of Waste." *Engineering and Science* 36, no. 4 (n.d.): 16–19, 30–32.

———. "Smell and Taste." *Scientific American* 186, no. 3 (June 1952): 28–32.

———. "Smog Control: Is It Just Around the Corner?" *Engineering and Science* (November 1962): 9–14.

———. "Smog Research Pays Off." *Engineering and Science* (May 1952): 11–16.

Haagen-Smit, Arie J., Margaret F. Brunelle and Jan W. Haagen-Smit. "Ozone Cracking in the Los Angeles Area." *Rubber Chemistry and Technology* 32, no. 4 (n.d.): 1134–42, 1959.

Hahn, Kenneth. In *Air Pollution—1967: Hearings Before the Subcommittee on Air and Water Pollution of the Committee on Public Works, United States Senate, 90th Congress, First Session on Problems and Progress Associated with Control of Automobile Exhaust Emissions, Part 1* (1967). Statement of Kenneth Hahn, Supervisor, Board of Supervisors, County of Los Angeles, State of California.

Heinen, Charles M., and Walter S. Fagley Jr. *Smog—The Learning Years—Building the 88th Story* (SAE Technical Paper Series 890813). Warrendale, PA: Society of Automotive Engineers, 1989.

Johnstone, H.F., Louis C. McCabe and M.D. Thomas. "Chemistry of Pollutants in the Atmosphere." *Industrial and Engineering Chemistry* 48, no. 9 (1956): 1483.

Kennedy, Harold W. "The History, Legal and Administrative Aspects of Air Pollution Control in the County of Los Angeles." Report to the Board of Supervisors, Los Angeles County, May 9, 1954.

Koster, Betty. "A History of Air Pollution Control Efforts in Los Angeles County." Los Angeles County Air Pollution Control District, August 31, 1956.

Kunkel, Harry E. "Air Pollution Control." Letter to C.L. Senn, September 28, 1945. Los Angeles City Archives Council Communications, Box A-832, vol. 3588, 15399.

Lillard, Richard G. *Eden in Jeopardy*. New York: Knopf, 1966.

Lodge, J.P. "A.J. Haagen-Smit" [obituary]. *Nature* 267 (n.d.): 565–66.

Los Angeles City Health Department. Progress Report on Smoke Elimination, August 25, 1943. Los Angeles City Archives Council Communications, Box A-832, vol. 3588, 15464.

Los Angeles County Air Pollution Control District. "Factory in the Sky." *LACAPCD Publication No. 41-A*, July 1954.

Los Angeles County Smoke and Fumes Commission. Report to the Los Angeles County Board of Supervisors, March 13, 1944. (CIMA0015 740/1526).

McCabe, Louis C. "Annual Report 1947–48." Los Angeles County Air Pollution Control District, 1949.

———. Letter to Los Angeles County Board of Supervisors, February 25, 1949.

———. Letter to Los Angeles County Board of Supervisors, September 13, 1948.

———. “National Trends in Air Pollution.” In *Proceedings of the First National Air Pollution Symposium*, 50–54. Los Angeles: Stanford Research Institute, 1949.

———. “Text of M’Cabe’s Report on Smog.” *Los Angeles Times*, September 15, 1948.

Mitchell, Daniel J.B. “Clearing the Air: What the Times Called for in 1947.” 2016. www.anderson.ucla.edu/documents/areas/fac/management/D.Mitchell_Clearing_the_Air_2016.pdf.

“Motor Vehicles Industry Efforts to Reduce Air Pollution from Exhaust.” Seattle, WA: Automobile Manufacturers Association, August 16, 1957.

Nicholson, Arnold. “Los Angeles Battles the Murk.” *Saturday Evening Post* 232, no. 25 (December 19, 1959): 17–19, 90–93.

Price Differential Between Leaded and Unleaded Gasoline: Hearing Before the Committee on Energy and Natural Resources. 95th Cong., S. Doc. No. 95-124 (1978).

Rosenberg, Jeremy. “How Los Angeles Began to Put Its Smoggy Days Behind.” February 13, 2021. www.kcet.org/history-society/how-los-angeles-began-to-put-its-smoggy-days-behind.

Royce, Stephen W. *A Life Remembered: The Memoirs of Stephen W. Royce.* Laguna Niguel, CA: Royal Literary Publications, 1984.

Schiller, Ronald. “The Los Angeles Smog.” *National Municipal Review* 44, no. 11 (n.d.): 558–64, 1955.

Senn, Charles L. Letter to George M. Uhl, July 30, 1943. Los Angeles City Archives Council Communications, Box A-832, Vol. 3588, 15464.

South Coast Air Quality Management District. “Historic Ozone Air Quality Trends.” November 3, 1996.

———. “Southland’s First Stage One Episode in Five Years Underscores Need to Accelerate Air Pollution Control Program.” Press release, July 11, 2003.

———. “Upland, Calif., Had Last Stage III Smog Alert in U.S.” News release, May 1997.

Sperling, Daniel, and Jennifer Dill. “Unleaded Gasoline in the United States: A Successful Model of System Innovation.” *Transportation Research Record* 1175 (n.d.): 45–46, 1988.

Stephens, Edgar R., Philip L. Hanst, Robert C. Doerr and William E. Scott. “Reactions of Nitrogen Dioxide and Organic Compounds in Air.” *Industrial and Engineering Chemistry* 48, no. 9 (1956): 1498–1504.

Tucker, Raymond R. “The Los Angeles Smog Report.” *Los Angeles Times*, January 19, 1947.

Wagner, Henry R. *Juan Rodriguez Cabrillo: Discoverer of the Coast of California.* San Francisco: California Historical Society, 1941.

Went, Frits W. "The General Problem of Air Pollution and Plants." *Proceedings of the First National Air Pollution Symposium, Pasadena, CA, November 10–11, 1949.*

Zeder, James C. "Million Dollar Search: What the Automotive Industry Is Doing to Help Los Angeles Fight Its Smog Problem." *Proceedings of the Third National Air Pollution Symposium, April 18–20, 1955.*

———. "Reduction of Exhaust Emission by Induction of System Devices." Quoted in *Motor Vehicle Industry Efforts to Reduce Air Pollution from Exhaust* (Seattle, WA: Automobile Manufacturers Association, August 16, 1957).

INDEX

E

F

G

H

J

K

L

M

N

O

P

R

S

T

U

V

W

X

Z

ABOUT THE AUTHOR

Carl Oliver is the author of *Plane Talk: Aviators and Astronauts Own Stories* and *Panama's Canal*. He coauthored *Business Ethics: The Path to Certainty*. His academic credentials include a BA in psychology from Stanford University, an MBA from California Lutheran University and both an MA in organization development and a PhD in human and organizational systems from Fielding Graduate University.

He began his professional career as a photojournalist in the Washington, D.C. area and served more than twenty years as a special agent for the U.S. Air Force and for another twenty years taught leadership, management, organization development and business ethics in industry and as a faculty member at Stanford, Loyola Marymount University–Los Angeles, the University of Redlands and California Lutheran University.

His research into moral reasoning has been published in the *Journal of Organizational Moral Psychology*, the *Journal of Adult Development*, the *International Journal of Leadership Studies*, online at the Markkula Center for Applied Ethics and online at the Dare Association. His writing on business ethics has appeared in *OD Practitioner*, *BizEd*, *Compliance & Ethics Professional*, *Ethikos* and *Computer Security Journal*.

Los Angeles' battle to defeat photochemical smog has provided Carl's undergraduate and master's degree students real-world demonstrations of the effects of stockholder and stakeholder theories; outcomes of corporate cultures that are company-centric or oriented toward charity, stewardship or citizenship; the role of continuous process improvement; and how to make any good organization or community even better.

Write to him at carl@ethicsprocess.com or PO Box 4888, Thousand Oaks, CA 91359-1888.